Titles in This Series

Volume

1 **Markov random fields and their applications,** Ross Kindermann and J. Laurie Snell

2 **Proceedings of the conference on integration, topology, and geometry in linear spaces,** William H. Graves, Editor

3 **The closed graph and P-closed graph properties in general topology,** T. R. Hamlett and L. L. Herrington

4 **Problems of elastic stability and vibrations,** Vadim Komkov, Editor

5 **Rational constructions of modules for simple Lie algebras,** George B. Seligman

6 **Umbral calculus and Hopf algebras,** Robert Morris, Editor

7 **Complex contour integral representation of cardinal spline functions,** Walter Schempp

8 **Ordered fields and real algebraic geometry,** D. W. Dubois and T. Recio, Editors

9 **Papers in algebra, analysis and statistics,** R. Lidl, Editor

10 **Operator algebras and K-theory,** Ronald G. Douglas and Claude Schochet, Editors

11 **Plane ellipticity and related problems,** Robert P. Gilbert, Editor

12 **Symposium on algebraic topology in honor of José Adem,** Samuel Gitler, Editor

13 **Algebraists' homage: Papers in ring theory and related topics,** S. A. Amitsur, D. J. Saltman, and G. B. Seligman, Editors

14 **Lectures on Nielsen fixed point theory,** Boju Jiang

15 **Advanced analytic number theory. Part I: Ramification theoretic methods,** Carlos J. Moreno

16 **Complex representations of GL(2, K) for finite fields K,** Ilya Piatetski-Shapiro

17 **Nonlinear partial differential equations,** Joel A. Smoller, Editor

18 **Fixed points and nonexpansive mappings,** Robert C. Sine, Editor

19 **Proceedings of the Northwestern homotopy theory conference,** Haynes R. Miller and Stewart B. Priddy, Editors

20 **Low dimensional topology,** Samuel J. Lomonaco, Jr., Editor

21 **Topological methods in nonlinear functional analysis,** S. P. Singh, S. Thomeier, and B. Watson, Editors

22 **Factorizations of $b^n \pm 1$, $b = 2$, 3, 5, 6, 7, 10, 11, 12 up to high powers,** John Brillhart, D. H. Lehmer, J. L. Selfridge, Bryant Tuckerman, and S. S. Wagstaff, Jr.

23 **Chapter 9 of Ramanujan's second notebook—Infinite series identities, transformations, and evaluations,** Bruce C. Berndt and Padmini T. Joshi

24 **Central extensions, Galois groups, and ideal class groups of number fields,** A. Fröhlich

25 **Value distribution theory and its applications,** Chung-Chun Yang, Editor

26 **Conference in modern analysis and probability,** Richard Beals, Anatole Beck, Alexandra Bellow, and Arshag Hajian, Editors

27 **Microlocal analysis,** M. Salah Baouendi, Richard Beals, and Linda Preiss Rothschild, Editors

28 **Fluids and plasmas: geometry and dynamics,** Jerrold E. Marsden, Editor

29 **Automated theorem proving,** W. W. Bledsoe and Donald Loveland, Editors

30 **Mathematical applications of category theory,** J. W. Gray, Editor

31 **Axiomatic set theory,** James E. Baumgartner, Donald A. Martin, and Saharon Shelah, Editors

32 **Proceedings of the conference on Banach algebras and several complex variables,** F. Greenleaf and D. Gulick, Editors

33 **Contributions to group theory,** Kenneth I. Appel, John G. Ratcliffe, and Paul E. Schupp, Editors

34 **Combinatorics and algebra,** Curtis Greene, Editor

Titles in This Series

Volume

Titles in This Series

Volume

68 **Differential geometry: The interface between pure and applied mathematics,** Mladen Luksic, Clyde Martin, and William Shadwick, Editors

69 **Methods and applications of mathematical logic,** Walter A. Carnielli and Luiz Paulo de Alcantara, Editors

70 **Index theory of elliptic operators, foliations, and operator algebras,** Jerome Kaminker, Kenneth C. Millett, and Claude Schochet, Editors

71 **Mathematics and general relativity,** James A. Isenberg, Editor

72 **Fixed point theory and its applications,** R. F. Brown, Editor

73 **Geometry of random motion,** Rick Durrett and Mark A. Pinsky, Editors

74 **Geometry of group representations,** William M. Goldman and Andy R. Magid, Editors

75 **The finite calculus associated with Bessel functions,** Frank M. Cholewinski

76 **The structure of finite algebras,** David C. Hobby and Ralph Mckenzie

77 **Number theory and its applications in China,** Wang Yuan, Yang Chung-chun, and Pan Chengbiao, Editors

78 **Braids,** Joan S. Birman and Anatoly Libgober, Editors

79 **Regular differential forms,** Ernst Kunz and Rolf Waldi

80 **Statistical inference from stochastic processes,** N. U. Prabhu, Editor

81 **Hamiltonian dynamical systems,** Kenneth R. Meyer and Donald G. Saari, Editors

82 **Classical groups and related topics,** Alexander J. Hahn, Donald G. James, and Zhe-xian Wan, Editors

83 **Algebraic K-theory and algebraic number theory,** Michael R. Stein and R. Keith Dennis, Editors

84 **Partition problems in topology,** Stevo Todorcevic

85 **Banach space theory,** Bor-Luh Lin, Editor

86 **Representation theory and number theory in connection with the local Langlands conjecture,** J. Ritter, Editor

87 **Abelian group theory,** Laszlo Fuchs, Rüdiger Göbel, and Phillip Schultz, Editors

88 **Invariant theory,** R. Fossum, W. Haboush, M. Hochster, and V. Lakshmibai, Editors

89 **Graphs and algorithms,** R. Bruce Richter, Editor

90 **Singularities,** Richard Randell, Editor

91 **Commutative harmonic analysis,** David Colella, Editor

92 **Categories in computer science and logic,** John W. Gray and Andre Scedrov, Editors

93 **Representation theory, group rings, and coding theory,** M. Isaacs, A. Lichtman, D. Passman, S. Sehgal, N. J. A. Sloane, and H. Zassenhaus, Editors

94 **Measure and measurable dynamics,** R. Daniel Mauldin, R. M. Shortt, and Cesar E. Silva, Editors

95 **Infinite algebraic extensions of finite fields,** Joel V. Brawley and George E. Schnibben

96 **Algebraic topology,** Mark Mahowald and Stewart Priddy, Editors

97 **Dynamics and control of multibody systems,** J. E. Marsden, P. S. Krishnaprasad, and J. C. Simo, Editors

98 **Every planar map is four colorable,** Kenneth Appel and Wolfgang Haken

99 **The connection between infinite dimensional and finite dimensional dynamical systems,** Basil Nicolaenko, Ciprian Foias, and Roger Temam, Editors

100 **Current progress in hyperbolic systems: Riemann problems and computations,** W. Brent Lindquist, Editor

101 **Recent developments in geometry,** S.-Y. Cheng, H. Choi, and Robert E. Greene, Editors

102 **Primes associated to an ideal,** Stephen McAdam

103 **Coloring theories,** Steve Fisk

Titles in This Series

Volume

104 Accessible categories: The foundations of categorical model theory, Michael Makkai and Robert Paré

105 Geometric and topological invariants of elliptic operators, Jerome Kaminker, Editor

106 Logic and computation, Wilfried Sieg, Editor

107 Harmonic analysis and partial differential equations, Mario Milman and Tomas Schonbek, Editors

108 Mathematics of nonlinear science, Melvyn S. Berger, Editor

Mathematics of
Nonlinear Science

CONTEMPORARY MATHEMATICS

108

Mathematics of Nonlinear Science

Proceedings of an AMS Special Session
held January 11–14, 1989

Melvyn S. Berger, Editor

AMERICAN MATHEMATICAL SOCIETY • PROVIDENCE, RHODE ISLAND

The AMS Special Session on Mathematics of Nonlinear Science was held at the 847th Meeting of the American Mathematical Society in Phoenix, Arizona, on January 11–14, 1989. Research was partially supported from an AFOSR grant and an NSF grant.

1980 *Mathematics Subject Classification* (1985 *Revision*). Primary 70-XX, 70DXX, 70FXX, 70KXX; Secondary 34BXX, 35-XX, 35BXX, 49-XX, 49FXX, 58-XX, 58BXX.

Library of Congress Cataloging-in-Publication Data

Mathematics of nonlinear science: proceedings of an AMS special session held January 11–14, 1989/Melvyn S. Berger, editor.
p. cm.—(Contemporary mathematics, ISSN 0271-4132; v. 108)
ISBN 0-8218-5114-4 (alk. paper)
1. Nonlinear theories—Congresses. 2. Dynamics—Congresses. 3. Differential equations—Congresses. 4. Differential equations, Partial—Congresses. I. Berger, Melvyn S. (Melvyn Stuart), 1939–. II. American Mathematical Society. Meeting (1989: Phoenix, Ariz.) III. Series.
QA427.M37 1990
515′.35—dc20
90-574
CIP

Printed in the United States of America.
The paper used in this book is acid-free and falls within the guidelines established to ensure permanence and durability. ♾
Portions of this publication were typeset using $\mathcal{AMS}$-TEX, the American Mathematical Society's TEX macro system.
10 9 8 7 6 5 4 3 2 1 95 94 93 92 91 90

Contents

Preface

The articles presented here were delivered as talks at the 1989 Annual Meeting of the American Mathematical Society in Phoenix, Arizona held during the month of January, 1989. The occasion for the delivery of the talks was a special session that I ran at the meeting entitled, "Mathematics of Nonlinear Science." The topic, Mathematics of Nonlinear Science, is a relatively new one. It distinguishes two types of mathematical areas in science: linear science and nonlinear science. Nonlinear science encompasses a large and rapidly growing area of new thinking concerning the relationship of mathematics to science where the fundamental laws of nature are extended beyond common sense into a very new area where the dual aspects of order and chaos bound. The papers presented here deal primarily with mathematical aspects of physical science and non-chaotic phenomena. In general the papers are analytic in nature and I hope have a certain topical unity. The material discussed from the mathematical side varies between ordinary and partial differential equations. Important new areas are discussed, such as instability, global extensions of the KAM theory, new ideas concerning integrable systems, bifurcation and its applications in fluids. Various aspects of gauge theory are also discussed. The list goes on and on. All in all the topics presented here present a good survey of some new ideas in the topics at hand. It is hoped that readers will be led to new vistas in this volume and may contribute new and exciting results to a fascinating subject area.

M. S. Berger

February 8, 1990

Contemporary Mathematics
Volume 108, 1990

Multiple steady states in tubular chemical reactors

ROGER K. ALEXANDER

Abstract. The steady-state temperature and reactant concentration in a tubular chemical reactor are solutions of a two-point boundary problem for two coupled nonlinear second-order differential equations. This paper describes the multiple solutions of this problem, which have been obtained by rigorous analysis, by formal asymptotic methods, and by numerical bifurcation analysis. When the activation energy and the heat transfer coefficient are large, nearly extinguished states exist with arbitrarily large multiplicity. The proof of this fact is outlined.

1. Introduction. This paper is an exposition of mathematical results on the pseudo-homogeneous axial dispersion model of a nonadiabatic tubular chemical reactor. We confine our attention to the steady states. This introductory section will summarize the older rigorous results and outline the picture provided by formal asymptotic methods and numerical bifurcation analysis. In the following section we sketch the proof that the model can have arbitrarily many steady states if the activation energy and the heat transfer coefficient are sufficiently large.

The steady states of the reactor are solutions of the following two-point boundary problem for the temperature T and reacting species concentration C. In these equations the reactor length has been normalized to unity. Constants H, M, B, D, β, γ are explained below. The equations are:

$$(1.1)\quad \begin{cases} \frac{1}{H}T'' - T' - \beta(T-1) + BDCe^{-\gamma/T} = 0, & 0 < x < 1; \\ T' - H(T-1) = 0, & x = 0; \\ T' = 0, & x = 1; \\ \frac{1}{M}C'' - C' - De^{-\gamma/T}C = 0, & 0 < x < 1; \\ C' - M(C-1) = 0, & x = 0; \\ C' = 0, & x = 1 \end{cases}$$

For a thorough discussion of the assumptions under which these equations are derived, consult the review in [22]. Here we only recapitulate these assumptions in brief: the reactor vessel is a long narrow tube, so that radial gradients of concentration and temperature are negligible; the chemistry is described by a single first-order irreversible exothermic reaction with Arrhenius temperature

The detailed version of this paper has been submitted for publication elsewhere.
1980 *Mathematics subject classifications.* 80A32, 34B15.

dependence; the reaction is superimposed on a mean flow which is fixed and constant; dispersion is by Taylor diffusion; finally, the reactor exchanges heat with a cooling jacket kept at a fixed (nondimensional) temperature $T = 1$.

According to these assumptions, we identify parameters in the Equations (1.1): M and H are the Peclet numbers for mass and heat, respectively, D is the Damköhler number, and B the heat release of the reaction. Our attention will be directed later on to the two remaining parameters: γ, the activation energy, and β, the coefficient of heat transfer. It has been known for a long time that the system (1.1) can have up to seven solutions, though only threefold multiplicity could be proved before now. In the second section of this paper we outline a proof that the system (1.1) can have arbitrarily many solutions when β and $\beta^{-1}\gamma$ are large enough. Full details will appear elsewhere [3].

First we shall describe what is known about the multiplicity of steady states. For more details refer to the author's survey [2]. A valuable recent survey from the point of view of chemical engineering is [15], where analytical criteria for uniqueness and stability of the steady state are also discussed.

We begin, as the mathematical analysts did, with the adiabatic ($\beta = 0$) reactor, assuming equal Peclet numbers ($H = M$). In this situation you can multiply the second differential equation by B and add it to the first to eliminate the nonlinear term. This yields a linear boundary problem for the quantity $T + BC$ (called in the combustion literature a Schwab-Zel'dovich variable), whose solution is elementary. This reduces the system (1.1) to a single nonlinear second-order boundary problem for T, say. This equation can be analyzed by phase-plane methods [13], or by monotone iteration methods based on the maximum principle [4], [12], [17].

These methods show that the adiabatic reactor with equal Peclet numbers can have up to three steady states. The monotone methods yield stability information as well: solutions are ordered; the largest and smallest are stable, the intermediate one unstable. Analytical criteria, in terms of the parameters in the equations, can be given for uniqueness of the steady state; when there is exactly one solution, it is globally asymptotically stable.

This reduction cannot be carried out if the reactor is nonadiabatic ($\beta > 0$). Analysis still yields up to three steady states, but there is no monotoneity in the full system (1.1): distinct solution profiles can cross each other [19], and a unique steady state need not be stable, for there can be Hopf bifurcation in the associated time-dependent problem [5], [6]. Moreover numerical computations exhibit steady state multiplicities up to five [9], [14], [20], [21]. Hard work with the maximum principle gives bounds on solutions [18].

Kapila and Poore clarified the situation in 1982 [11]. Exploiting the assumption of large activation energy, they used the method of matched asymptotic expansions to construct formal solutions of Eq. (1.1), exhibiting seven distinct types. A crucial concept in their construction is that of a reaction zone: a narrow region of x within which C and T vary rapidly ("active chemistry"). Outside such a zone the nonlinear terms in Eq. (1.1) do not contribute significantly ("frozen chemistry"), and T and C are governed by linear equations.

A formal solution constructed this way is said to exhibit low, intermediate, or high conversion in case the reactant concentration at the outlet ($C(1)$) is nearly

1, well between 0 and 1, or nearly 0, respectively. "Nearly" is measured in units of γ^{-1}. An intermediate- or high-conversion steady state has one reaction zone, which can appear near the reactor inlet $\left(x = O(\gamma^{-1})\right)$, well in the interior, or near the outlet $\left(x = 1 - O(\gamma^{-1})\right)$. This accounts for six distinct solutions. Some of these states are realizable only if the heat transfer coefficient β is large enough. Finally, there is the low-conversion ("extinguished") steady state, in which T and C remain close to their inlet values throughout the reactor tube.

This analysis is "merely" formal, but it has been confirmed by numerical bifurcation analysis [7], [8]. Moreover, the qualitative picture of solution profiles that it provides helps in devising efficient numerical schemes to trace parametric solution branches [10].

In the remainder of this paper we forsake all the solutions exhibiting reaction zones, and regard only the low-conversion steady states. We show that these may be obtained in any multiplicity desired by making β and γ/β large. In these steady states the temperature and concentration vary from their inlet values by only $O(\gamma^{-1}\log\beta)$ and $O(\gamma^{-1}\beta)$ respectively; the temperature is oscillatory in space with amplitude growing slowly as the reactor is traversed from inlet to outlet.

The mechanism producing these solutions was discovered in [1]. Sharper multiplicity estimates, and proof that the mechanism yields solutions of Eq. (1.1) will appear in [3].

It is interesting to note that arbitrary multiplicity of solutions has also been discovered in the somewhat related problem of the porous catalyst pellet [23]. In this problem too a rigorous analysis seems difficult. In [24] the shooting method is used to reduce the problem to a finite dimensional system of equations. These equations are analyzed by the methods of singularity theory: quantities which appear in the analysis are defined in terms of solutions of certain differential equation initial value problems, and are computed by integrating the differential equations numerically.

2. The Low-Conversion Steady States. This section is devoted to a study of those solutions of Eq (1.1) in which T and C remain close to 1. Here we explain the main steps in the proof; detailed estimates will appear elsewhere [3].

We use the activation energy γ, assumed large, as a microscope with which to view the nearly extinguished steady states. Make the change of variables (and parameter) in Eq. (1.1),

$$\begin{aligned} T &= 1 + \gamma^{-1} y, \\ C &= 1 + \gamma^{-1} z, \\ D &= \gamma^{-1} \lambda e^{\gamma}, \end{aligned}$$

to obtain an equivalent boundary problem for (y, z):

$$\text{(2.1a)} \qquad \frac{1}{H}y'' - y' - \beta y + B\lambda(1+\gamma^{-1}z) \cdot \exp\left[y/(1+\gamma^{-1}y)\right] = 0, \qquad 0 < x < 1,$$

$$\text{(2.1b)} \qquad y' - Hy = 0, \qquad x = 0,$$

$$\text{(2.1c)} \qquad y' = 0, \qquad x = 1,$$

$$\text{(2.1d)} \qquad \frac{1}{M}z'' - z' - \gamma^{-1}\lambda \exp\left[y/(1+\gamma^{-1}y)\right] z = \lambda \exp\left[y/(1+\gamma^{-1}y)\right], \qquad 0 < x < 1,$$

$$\text{(2.1e)} \qquad z' - Mz = 0, \qquad x = 0,$$

$$\text{(2.1f)} \qquad z' = 0, \qquad x = 1.$$

In this system z should not influence y very much if z is not too large, because z is multiplied by γ^{-1} in Eq. (2.1a). On the other hand it is not hard to show that Eqs. (2.1d-f) would determine z if y were known. Our method for solving Eq. (2.1) consists of two steps. First we ignore z in Eq. (2.1a); that is, we consider

$$\text{(2.2)} \qquad \frac{1}{H}y'' - y' - \beta y + B\lambda \exp\left[\frac{y}{1+\gamma^{-1}y}\right] = 0, \qquad 0 < x < 1,$$

subject to the boundary conditions (2.1b,c). We prove the desired multiplicity result for this problem, together with useful bounds on the solutions. Then we use the solutions of Eq. (2.2) satisfying the first boundary condition (2.1b) as starting points for an iteration—convergent when γ^{-1} is small—to obtain solutions of the full problem (2.1).

Bear in mind that Eq. (2.1) is derived from Eq. (1.1) by a nonsingular change of variables and so is equivalent to it. (It is not the first term in an expansion in inverse powers of γ.)

We take the constants M, H, B, λ to be fixed. Theorem 1 states our multiplicity result.

THEOREM 1. *Let $n > 0$ be an integer. Then the boundary problem (2.2), (2.1b,c) has at least n solutions when β and γ/β are sufficiently large. Moreover the solutions are nonnegative and satisfy*

$$\max_{0\le x\le 1} y(x) \le \text{const} \log \beta$$

with a constant independent of β and γ.

We denote by Y the upper bound for the solutions.

We shall actually use the spatially oscillatory nature of these solutions to construct the steady states of the full problem (2.1). To see the mechanism giving multiple oscillatory solutions we look in the phase plane. In Eq. (2.2) let

$$y_1 = y,$$
$$y_2 = y'.$$

Then Eq. (2.2) is

(2.3)
$$y_1' = y_2$$
$$y_2' = H\left\{y_2 + \beta y_1 - B\lambda \exp\left[\frac{y_1}{1+\gamma^{-1}y_1}\right]\right\},$$

and the boundary conditions (2.1b,c) are

(2.4a)
$$y_2(0) - Hy_1(0) = 0,$$

(2.4b)
$$y_2(1) = 0.$$

In the phase plane, then, a solution of the two-point boundary problem (2.2), (2.1b,c) is represented by a trajectory of Eq. (2.3) departing from the "line of initial conditions" $y_2 = Hy_1$ at $x = 0$ and reaching the horizontal axis $y_2 = 0$ when $x = 1$.

The system (2.3) has three critical points when $\beta/B\lambda > e$ and γ/β is large. We designate them by $(\alpha_j, 0)$, $j = 0, 1, 2$ and $0 < \alpha_0 < \alpha_1 < \alpha_2$.

The first and third are saddles, given approximately by

$$\alpha_0 = B\lambda/\beta + O(\beta^{-2})$$
$$\alpha_2 = (B\lambda/\beta)e^{\gamma}\left(1 + O(\gamma^2 e^{-\gamma})\right).$$

Note that with β and γ large, $(\alpha_0, 0)$ is close to the origin; $(\alpha_2, 0)$ is so far away that it is irrelevant to our analysis. The pivotal point is the middle one; we write simply α for α_1:

(2.5)
$$\alpha = \log(\beta/B\lambda) + \log\log(\beta/B\lambda) + o(1).$$

This point is an unstable spiral provided that

$$\frac{\alpha}{(1+\gamma^{-1}\alpha)^2} > 1 + \frac{H}{4\beta}$$

holds, which because of Eq. (2.5) is true when β is large enough. Then the complex eigenvalues of the linearization of Eq. (2.3) around $(\alpha, 0)$ are $H/2 \pm i\omega$, with

$$\omega^2 = H\beta\left(\frac{\alpha}{(1+\gamma^{-1}\alpha)^2} - 1\right) - H^2/4.$$

By Eq. (2.5) we have

(2.6)
$$\omega \sim (H\beta \log \beta)^{1/2}$$

for large β.

To exhibit the mechanism producing multiple solutions, we express the boundary condition (2.1c) in terms of the polar angle at $(\alpha, 0)$. By $y(x;\eta) = (y_1(x;\eta), y_2(x;\eta))^T$ we mean the solution of Eq. (2.3) satisfying the initial condition

$$y_1(0) = \eta, \qquad y_2(0) = H\eta. \tag{2.7}$$

Then $y(\cdot;\eta)$ is a solution of the boundary problem (2.3), (2.4a,b) if $y_2(1;\eta) = 0$ ("shooting method").

Now use the rule

$$\widehat{\theta}(x;\eta) = \arctan \frac{y_2(x;\eta)}{y_1(x;\eta) - \alpha}$$

to determine $\widehat{\theta}$ depending continuously on both variables $x \geq 0$, $\eta \geq 0$, understanding that

$$\pi \geq \widehat{\theta}(0;\eta) \geq 0.$$

Solutions of our boundary problem correspond to values of η for which

$$\widehat{\theta}(1;\eta) = k\pi, \qquad k \in \mathbb{Z}. \tag{2.8}$$

Refer now to the phase portrait, Fig. 1. It is easy to see that

$$\widehat{\theta}(1;0) > \pi. \tag{2.9}$$

Next we find an abscissa to the right of α,

$$Y = \text{const} \ \log \beta$$

for which

$$\widehat{\theta}(1;Y) > -\pi. \tag{2.10}$$

Finally we show that there is an η^* with $0 < \eta^* < Y$ such that

$$\widehat{\theta}(1;\eta^*) = -\omega + O(1). \tag{2.11}$$

The proof of Theorem 1 now follows by combining Eqs. (2.9), (2.10), (2.11): from Eq. (2.6), ω is as large as we please. Continuous dependence and the intermediate value theorem yield, according to Eq. (2.8), $[\omega/\pi] + 1$ solutions of our boundary problem with $0 < \eta < \eta^*$, and again $[\omega/\pi] - 1$ solutions with $\eta^* < \eta < Y$.

To conclude the discussion of Theorem 1 we indicate briefly how to establish Eqs. (2.10) and (2.11).

Consider the conservative system

$$\begin{aligned} y_1' &= y_2 \\ y_2' &= H\beta\left(y_1 - \frac{B\lambda}{\beta} \exp\left[\frac{y_1}{1 + \gamma^{-1} y_1}\right]\right), \end{aligned} \tag{2.12}$$

obtained by deleting y_2 from the right side of the second equation in Eq. (2.3). The Hamiltonian for Eq. (2.12) turns out to be a Lyapunov function for Eq. (2.3) run backward in x. It follows that the right branch of the stable manifold of the saddle $(\alpha_0, 0)$ for Eq. (2.3) falls into the spiral as $x \to -\infty$, and we may choose any abscissa for Y such that $Y > \alpha$ and the point (Y, HY) is to the right of the rightmost intersection of the stable manifold of the saddle with the line of initial conditions.

To find η^* for Eq. (2.11), we make an affine change of variable in Eq. (2.3) to bring the spiral point $(\alpha, 0)$ to the origin and put the linear part in canonical form. The transformed system is

$$u' = Au + \omega N(u_1) \begin{bmatrix} 0 \\ 1 \end{bmatrix}, \tag{2.13}$$

with $u = (u_1, u_2)^T$, N a nonlinear function satisfying $N(0) = N'(0) = 0$, $N''(0) \sim 1$, and

$$A = \begin{bmatrix} H/2 & \omega \\ -\omega & H/2 \end{bmatrix}. \tag{2.14}$$

The boundary conditions (2.4a,b), expressed in terms of the new variables, are

$$u_2(0) = \frac{H}{2\omega} u_1(0) + \frac{H\alpha}{\omega}, \tag{2.15a}$$

$$u_2(1) = -\frac{H}{2\omega} u_1(1). \tag{2.15b}$$

So the line of initial conditions, Eq. (2.15a) passes close to the origin in the (u_1, u_2) plane, by Eqs. (2.5), (2.6). Consider the solution of Eq. (2.13) with $u_1(0) = 0$, $u_2(0) = H\alpha/\omega$ on the initial line. Because this point is close to the origin we expect its trajectory to be near the solution of the linearized equation $u' = Au$ with the same initial data. This is indeed the case, but the proof is slightly delicate because of the large coefficient ω in the nonlinear term in Eq. (2.13). Solutions of $u' = Au$ wind rapidly around the spiral point (angular velocity ω). A careful application of the method of averaging [16] yields the desired result. Details are given in [3].

We now use the conclusions of Theorem 1 about multiplicity of solutions and bounds on their magnitude, together with phase plane information, to obtain solutions of Eq. (2.1). The key to our result is the following lemma.

LEMMA. *There is a constant C not depending on β or γ, and for each β a constant $K = K(\beta)$ not depending on γ, such that for every η with $0 < \eta < Y$ (Y given in Eq. 2.10) there is a unique ζ with $0 < \zeta < C\lambda e^Y$ for which the solution of the differential equations (2.1a), (2.1d) with initial conditions*

$$y(0) = \eta, \qquad y'(0) = H\eta, \tag{2.16}$$

$$z(0) = \zeta, \qquad z'(0) = M\zeta, \tag{2.17}$$

satisfies $z'(1) = 0$. *Moreover, if* $\tilde{y}$ *is the solution of Eq. (2.2) with the same initial conditions (2.16), the estimates*

$$\max_{0\le x\le 1} |y(x) - \tilde{y}(x)| \le \gamma^{-1} K(\beta)$$
$$\max_{0\le x\le 1} |y'(x) - \tilde{y}'(x)| \le \gamma^{-1} K(\beta) \tag{2.18}$$

hold.

We prove the Lemma by setting up a convergent iteration. Let $z_0(x) = 0$, $0 \le x \le 1$ and for $k = 1, 2, \ldots$ define functions (y_k, z_k) by

$$y_k := \text{solution of Eq. (2.1a), (2.16), using } z = z_{k-1}.$$
$$z_k := \text{solution of Eq. (2.1d,e,f), using } y = y_k.$$

The boundary problem for z_k is easily shown to have a unique solution. Then simple differential inequalities, based on the fact that z appears with coefficient γ^{-1} in Eq. (2.1a), establish convergence of the sequence, and the estimate (2.18).

With these ingredients, Theorem 2 can now be proven.

THEOREM 2. *The boundary problem (1.1) has solutions in the form of low- conversion steady states; their multiplicity may be made to exceed any preassigned number by making* β *sufficiently large provided* $\gamma^{-1}\beta$ *remains small.*

To prove Theorem 2 it suffices to establish Eqs. (2.9), (2.10), (2.11) with $\widehat{\theta}$ defined using $(y_1, y_2) = (y, y')$ from the Lemma. The estimate (2.18) shows that $\widehat{\theta}$ is well-defined, and that the solution y of the (y, z) system is C'-close to the solution $\tilde{y}$ of Eq. (2.2) that has the same initial data. Thus Eqs. (2.9-11) hold for the y-components of solutions of the system as well, and the proof is completed as before.

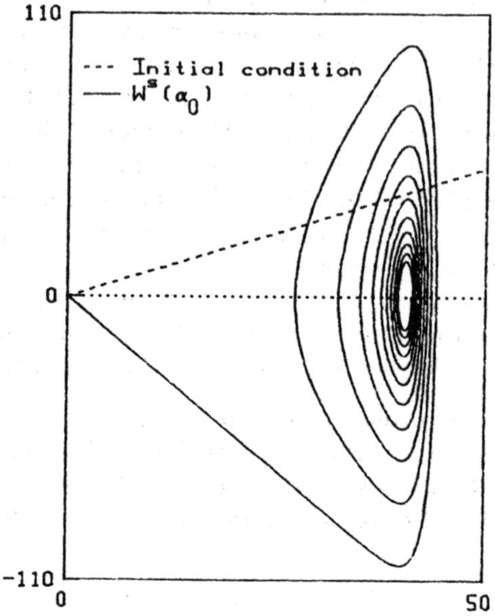

Fig. 1: β=10, γ=200, H=1, Bλ=10^{-12}.

REFERENCES

1. R. Alexander, *The multiple steady states of a nonadiabatic tubular reactor*, J. Math. Anal. Appl. **101** (1984), 12–22.

2. __________, *Mathematical analysis of tubular reactors*, in "Reacting Flows: Combustion and Chemical Reactors," ed. G.S.S. Ludford, American Mathematical Society, Providence, Rhode Island, 1986. vol 2, pp. 317–330.

3. __________, *Spatially oscillatory steady states of tubular chemical reactors*, SIAM J. Math. Analysis, to appear.

4. D. S. Cohen, *Multiple stable solutions of nonlinear boundary value problems arising in chemical reactor theory*, SIAM J. Appl. Math. **20** (1971), 1–13.

5. __________, *Multiple solutions and periodic oscillations in nonlinear diffusion processes*, SIAM J. Appl. Math. **25** (1973), 640–654.

6. __________ and A. B. Poore, *Tubular chemical reactors; the 'lumping approximation' and bifurcation of oscillatory states*, SIAM J. Appl. Math. **27** (1974), 416–429.

7. R. F. Heinemann and A. B. Poore, *Multiplicity and oscillatory dynamics of the tubular reactor*, Chem. Engrg. Sci. **36** (1981), 1411.

8. __________ and __________, *The effect of activation energy on tubular reactor multiplicity*, Chem. Engrg. Sci. **37** (1982), 128–130.

9. V. Hlaváček, H. Hofmann and M. Kubíček, *Transient axial heat and mass transfer in tubular reactors. The stability considerations-II*, Chem. Engrg. Sci. **26** (1971), 1629–1634.

10. K. F. Jensen and W. H. Ray, *The bifurcation behavior of tubular reactors*, Chem. Engrg. Sci. **37** (1982), 199–238.

11. A. K. Kapila and A. B. Poore, *The steady response of a nonadiabatic tubular reactor: new multiplicities*, Chem. Engrg. Sci. **37** (1982), 57–76.

12. H. B. Keller, *Existence theory for multiple solutions of a singular perturbation problem*, SIAM J. Math. Anal. **3** (1972), 86–92.

13. L. Markus and N. Amundson, *Nonlinear boundary value problems arising in chemical reactor theory*, J. Diff. Eq. **4** (1968), 102–113.

14. C. R. McGowin and D. D. Perlmutter, *Tubular reactor steady state and stability characteristics*, A. I. Ch. E. J. **17** (1971), 831–849.

15. M. Morbidella, A. Varma and R. Aris, *Reactor steady-state multiplicity and stability*, in "Chemical Reaction and Reactor Engineering," J.J. Carberry and A. Varma, eds., Marcel Dekker, New York, 1986, pp. 973–1054.

16. J. A. Sanders and F. Verhulst, "Averaging Methods in Nonlinear Dynamical Systems," Springer, New York, 1985.

17. D. H. Sattinger, *Monotone methods in nonlinear elliptic and parabolic boundary value problems*, Indiana Univ. Math. J. **21** (1972), 979–1000.

18. A. Varma, *Bounds on the steady state concentration and temperature in a tubular reactor*, Can. J. Chem. Engrg. **55** (1977), 629–632.

19. __________ and N. R. Amundson, *Some problems concerning the non-adiabatic tubular reactor. A priori bounds, qualitative behavior, preliminary uniqueness and stability considerations*, Can. J. Chem. Engrg. **50** (1972), 470–485.

20. __________ and __________, *Some observations on the uniqueness and multiplicity of steady states in non- adiabatic chemically reacting systems*, Can. J. Chem. Engrg. **51** (1973), 206–226.

21. __________ and __________, *The non-adiabatic tubular reactor: stability considerations*, Can. J. Chem. Engrg. **51** (1973), 459–467.

22. __________ and R. Aris, *Stirred pots and empty tubes*, in "Chemical Reactor Theory: A Review," ed. L. Lapidus and N.R. Amundson, Prentice-Hall, Englewood Cliffs, New Jersey, 1977, pp. 79–155.

23. J. M. Vega and A. Liñan, *Singular Langmuir-Hinshelwood reaction-diffusion problems: strong adsorption under quasi-isothermal conditions*, SIAM J. Appl. Math. **42** (1982), 1047–1068.

24. G. W. Witmer, V. Balakotaiah and D. Luss, *Multiplicity features of distributed systems - I. Langmuir-Hinshelwood reaction in a porous catalyst*, Chem. Engrg. Sci., **41** No. 1 (1986), 179–186.

Department of Mathematics, Iowa State University, Ames, Iowa 50011

Contemporary Mathematics
Volume **108**, 1990

Two New Approaches to Large Amplitude Quasi-Periodic Motions of Certain Nonlinear Hamiltonian Systems

M. S. Berger*

Center for Applied Mathematics
University of Massachusetts, Amherst

In this paper I wish to report on two different aspects of a key problem in nonlinear dynamics. The problem concerns large amplitude nonlinear quasi-periodic oscillations of certain simple conservative dynamical systems of the second order. In terms of differential equations, an example of the problem considered can be written

$$(1) \qquad \ddot{q} + \alpha q + \beta q^3 = h(t) \qquad \alpha, \beta \text{ constants}$$

Here $h(t)$ is a given quasi-periodic forcing term with fixed frequencies $\omega_1, \omega_2, \ldots \omega_n$, and possibly of large amplitude. The problem to be discussed concerns the possibility of finding quasi-periodic solutions of (1) and if so, determining the frequencies of these motions. In particular one asks if the frequencies of the solution $q(t)$ are identical with those of $h(t)$.

Section 1. Historical Background

This problem is a historic one dating back to Helmholtz's discussion of combination tones in his grand nineteenth century

1980 *Mathematics Subject Classification* (1985 *Revision*). 58A40, 58B15, 70H30, 70K40.
The final (detailed) version of this paper will be submitted for publication elsewhere.
*Research was partially supported by grants from the AFOSR and the NSF.

0271-4132/90 $1.00 + $.25 per page

treatise Sensations of Tone. It is known that if (1) is altered by adding a dissipation term (i.e., inserting the linear term $k\dot{q}$ with k a nonzero constant), then the altered equation (1) does generally possess quasi-periodic motions with frequencies exactly equal to the frequencies of $h(t)$ (see the appendix of J.J. Stoker's book Nonlinear Vibrations). For $k = 0$ nothing was known until 1965. Then J. Moser, in a paper published in the Communications of Pure and Applied Mathematics, studied (1) for small amplitude oscillations (i.e. $|h(t)|$ small) using the KAM theory for small Hamiltonian perturbations of an integrable Hamiltonian system. This approach does succeed in finding quasi-periodic solutions of (1) of small amplitude provided the constants α and β of (1) satisfy certain irrationality conditions and exclude certain degeneracies. However for large amplitude $h(t)$ nothing seems known. Moreover certain formal perturbation expansions for quasi-periodic solutions of (1) are known to diverge.

The new idea in this paper is easily summarized. I wish to find an approach to studying (1) and related systems utilizing the calculus of variations and Hilbert space arguments. This approach was crucial in finding new global ideas in the periodic case. To date, there has been no attempt to extend these ideas to the quasi-periodic case. This extension is the purpose of this article.

Section 2. An Approach by Nonlinear Partial Differential Equations and Optimization

Research discussed in this section represents joint work with Alex Eydeland. Our idea is to write down a standard nonlinear differential second-order equation that may have quasi-periodic solutions. We then find from this nonlinear ordinary differential equation the associated nonlinear partial differential equation whose solution will yield the function of n variables that we seek. Because the system is Hamiltonian we are then able to find a variational principle for this function of n variables. As a final step, we show that in certain cases this function of n variables can be determined by infinite-dimensional minimization techniques. Here is a case in point.

We consider the nonlinear ordinary differential equation

(1') $\ddot{q} - aq - bq^3 = h(t)$ where a, b are positive numbers

and $h(t)$ is a given quasi-periodic function of frequencies $(\omega_1, \omega_2, \ldots \omega_n)$. We seek a variational principle that determines a quasi-periodic solution $q(t)$ of frequencies $(\omega_1, \omega_2, \ldots \omega_n)$ for (1') assuming $h(t)$ is given as above (i.e., we seek a quasi-periodic solution $q(t)$ whose frequencies are identical with those of the forcing term $h(t)$). In symbols, this means we seek a function $u(t_1, t_2, \ldots, t_n)$ of n variables 2π-periodic in each t_i $(i = 1, 2, \ldots, n)$ with

(2') $$q(t) = u(\omega_1 t, \omega_2 t, \ldots \omega_n t).$$

We now derive a relationship between the functions $q(t)$ and $u(t_1, t_2, \ldots, t_n)$. Differentiating (2') with respect to t, we find

(3') $$\dot{q}(t) = \sum_{t=1}^{n} \omega_i u_i(\omega_1 t, \omega_2 t, \ldots, \omega_n t) \quad \text{where } u_i = u_{t_i}.$$

Differentiating once again, we find

(4') $$\ddot{q}(t) = \sum_{i,j=1}^{n} \omega_i \omega_j u_{t_i t_j}$$

$$= Lu.$$

Here Lu denotes the second-order differential operator defined by the right-hand side of (4'). This operator L is positive semidefinite when regarded as a second-order differential operator on R^n. Indeed, L is the square of any operator M, i.e.,

$$Lu = \sum_{i,j=1}^{n} \omega_i \omega_j u_{ij} = \left(\sum_{i=1}^{n} \omega_i u_i \right)^2 = (Mu)^2$$

where $Mu = \Sigma_{i=1}^{n} \omega_i u_{t_i}$. The characteristic form of L, $L_0(k)$, can be written in terms of the vector $k = (k_1, k_2, \ldots, k_n)$ as

$$L_0(k) = \left(\sum_{t=1}^{n} \omega_i k_i\right)^2$$

This form is semidefinite (for $n > 1$), since $L_0(\xi) = 0$ whenever we have $\Sigma_{i=1}^{n} \omega_i \xi_i = 0$ (i.e., on a hyperplane of codimension 1).

Consider now the following partial differential equation associated with (1'):

(5') $Lu - au - bu^3 = H(t_1, t_2, \ldots, t_n)$ (a, b positive constants),

defined on a torus $T^n = [-\pi, \pi]^n$ with opposite sides on each interval identified. Thus we seek solutions of (5') 2π periodic in each variable t_i $(i = 1, 2, \ldots n)$ and consider the functional on the Sobolev space H of functions in $H = W_{1,2}([-\pi, \pi]^n)$

$$(6') \quad \phi(u) = \int_{-\pi}^{\pi} \cdots \int_{-\pi}^{\pi} \cdots \int_{-\pi}^{\pi} \{(Mu)^2 + au^2 + \frac{1}{2} bu^4 + 2Hu\} dt_1 dt_2 \ldots dt_n$$

Lemma 1 If $u(x_1, x_2, \ldots, x_n)$ is a smooth critical point of the functional $\phi(u)$ on H, then the function

$$q(t) = u(\omega_1 t, \omega_2 t, \ldots \omega_n t)$$

is a solution of the problem (1). Moreover, $\phi(u)$ is convex on H.

Proof Indeed, $u_1(t_1, t_2, \ldots, t_n)$ must satisfy the Euler-Lagrange equation of $\phi(u)$. Thus, $q(t)$, defined by (2'), must satisfy (1'). The desired convexity follows immediately from computing the appropriate second derivative of $\phi(u)$.

Lemma 2 The functional (6') defined above attains its minimum in the space K of odd functions in H.

Proof We can prove that for $u \in K$,

(7') $$\int_{-\pi}^{\pi}\int_{-\pi}^{\pi}\int_{-\pi}^{\pi}\int_{-\pi}^{\pi} (Mu)^2 \geq \alpha\|u\|_H^2 \quad (\alpha \text{ a positive constant}).$$

To prove (7') over K we merely note the crossproduct term

$$\int_{-\pi}^{\pi}\int_{-\pi}^{\pi}\cdots\int_{-\pi}^{\pi} u_{t_i} u_{t_j} = 0 \qquad \text{for } i \neq j$$

This follows simply via integration by parts using parity considerations of elements $u \in K$.

As $a, b > 0$, $\inf_K \phi(u) > -\infty$. Moreover, the coerciveness and lower semicontinuity properties in H prevail by general functional analysis considerations.

However to date, because of lack of coerciveness, we have not firmly established the existence of absolute minimum for the functional (6) by this direct method of the calculus of variations. Thus we pass here to a second method, in which we have been successful in establishing the existence of a quasi-periodic smooth solution of the equation(1')for arbitrary quasi-periodic h(t).

Section 3. An Approach by Almost Periodic Functions

Harald Bohr initiated the theory of almost periodic functions in the 1920's as a generalization of the notion of quasi-periodic. An almost periodic continuous function can be simply described as the uniform limit of trigonometric polynomials. In order to study equation (1)

(1) $$\ddot{q} + \alpha q + \beta q^3 = h(t) \qquad \alpha, \beta \text{ constants}$$

we generalize the driving function h(t) to be simply a continuous almost periodic function in the sense of Bohr. One can then attempt to use the theory of almost periodic functions, developed in subsequent years by many researchers, to study the almost periodic solutions of (1), and to study exactly how the frequencies of the almost periodic function $q(t)$ reflects the frequencies of the forcing term $h(t)$. In the simplest case, when $h(t)$ is periodic, it is known for appropriate signs of the constants α and β, that the equation (1) has a unique periodic solution with the exact minimal period as $h(t)$. However, when we pass to this generalized, almost periodic formulation, and we consider $h(t)$ of large amplitude, then the frequencies of the solution $q(t)$ are not known. Certain facts about (1) can be ascertained, however, if we choose the constants α and β to be strictly negative. Ignoring the question of the frequencies of $h(t)$, one can study the problem (1) on the finite interval $[-T, T]$ in terms of the minimization of a convex functional $I(q)$ in the Sobolev space $W_{1,2}[+T, -T]$ with the norm

$$\| q \|^2 = \int_{-T}^{T} (|\dot{q}|^2 + |q|^2)\, dt$$

One obtains in this way a smooth solution of (1) $q(t)$ satisfying the boundary conditions

$$\dot{q}(-T) = \dot{q}(t) = 0$$

One then can extend the function so obtained $q(t)$ to the entire infinite interval $[-\infty, +\infty]$ in a periodic way. To get an almost periodic solution of the equation (1) on the entire interval, one chooses an appropriate subsequence of the extended functions $q(t)$ so obtained. The virtue of choosing the negative signs for the constants α and β are that in this case the associated functional

$$I(q) = \lim_{T\to\infty} \frac{1}{2T} \int_{-T}^{T} \left[|\dot{q}|^2 - \alpha |q|^2 - \frac{\beta}{2} |q|^4 + h(t) q \right] dt$$

is strictly convex as can be ascertained by simply checking the second variation of the functional $I(q)$ in any direction v in the Sobolev space. Moreover, the direct method of the calculus of variations insures that the absolute minimum of this functional is obtained at a function $\overline{q}$ of T as is clear from the strict convexity of the functional I. Furthermore, the associated minimum is clearly unique, and smooth by a simple application of the simple regularity theory for such problems.

It turns out that the extended function $\overline{q}(t)$ on the infinite interval can be shown to be almost periodic in the sense of Harald Bohr, in that such an almost periodic solution of the equation (1) is unique. One then has the final problem of deciding exactly what are the frequencies of this unique almost periodic function. It turns out that the notion of translation number of a continuous almost periodic function becomes crucial at this stage; and in fact, one can prove the following inequality for the pointwise behavior of the solution $q(t)$ relative to the almost periodic forcing term $h(t)$.

Suppose one is given an almost periodic forcing term $h(t)$ satisfying the Duffing equation (1') i.e.

(*)

$$\ddot{x} - ax - bx^3 = h(t)$$

where a and b are positive constants.

Question: Does this equation have an almost periodic solution with the same Fourier series frequencies as $h(t)$?

Very relevant points in answering this question are:

a) The equation (*) possesses a uniformly almost periodic solution $x(t)$.

b) The uniformly almost periodic solution of equation (*) is unique in the class of uniformly almost periodic functions. .

c) Any uniformly almost periodic solution $x(t)$ of equation (*) satisfies the following estimate:

$$\sup_{t\in\mathfrak{R}^1} |x(t)| \leq C \sup_{t\in\mathfrak{R}^1} |h(t)|$$

where C is an absolute constant.

These results allow the Fourier exponents of the solution $x(t)$ to be connected with the Fourier exponents of the forcing term $h(t)$ in a very direct manner. In particular, if $h(t)$ is quasiperiodic with N exponents independent over the rationals, one can deduce that the unique solution $x(t)$ is also quasiperiodic with the same Fourier exponents as $h(t)$. We state our result as follows:

THEOREM. The equation (1') whose forcing term $h(t)$ is any given quasi-periodic smooth function with arbitrary N frequencies independent over rational numbers, has one and only one smooth quasi-periodic solution $x(t)$. The frequencies of the solution $x(t)$ are exactly those of the forcing term $h(t)$.

The proof of this result will not be given here, but will be published elsewhere along with an extension to coupled systems of second order Hamiltonian systems with a convexity property extending (1').

An interesting research problem occurs when the signs of the constants a and b are allowed to vary. Then the minimization techniques discussed here must be supplemented to consider critical points of saddle point type. Our research on this topic, at this moment, is still in progress. An additional problem occurs for the analogous systems.

Contemporary Mathematics
Volume **108**, 1990

VORTICES FOR THE GINZBURG–LANDAU EQUATIONS —THE NONSYMMETRIC CASE IN BOUNDED DOMAIN

Y. Y. Chen

ABSTRACT. We discuss some results on the solutions of the steady–state Ginzburg–Landau equations with arbitrary parameter $\lambda > 0$ in two dimensional Euclidean space. In particular, we show that by means of the Cronström gauge condition and the constrained variational principle, there exist infinitely many nonsymmetric solutions on bounded domain Ω in $\mathbb{R}^2$.

1. INTRODUCTION.

The Ginzburg–Landau equations were formulated in 1950, which describe certain macroscopic effects in superconductivity [8]. It has been proven that this system of nonlinear partial differential equations is very useful in predicting the physical phenomena observed [1]. In the two dimensional case, the Ginzburg–Landau equations to the equilibrium states involve a complex valued function $\varphi = \varphi_1 + i\varphi_2$ and a two–component real valued vector $A = (A_1, A_2)$, which is the vector potential of the magnetic field. The equations can be written [9]

$$\text{(1a)} \qquad \Sigma_{k=1}^{2} (\partial_k - iA_k)^2 \varphi = \tfrac{\lambda}{2}(|\varphi|^2 - 1)\varphi$$

$$\text{(1b)} \qquad \partial_k(\partial_k A_j - \partial_j A_k) = \mathrm{Im}[\varphi(\partial_j + iA_j)\overline{\varphi}] \qquad k,j = 1,2, \quad k \neq j$$

where λ is a positive numerical parameter depending on the superconducting material, $0 < \lambda < 1$ describes type I superconductor, while $\lambda > 1$ describes type II. The state where $|\varphi|^2 \cong 0$ behaves normally, and the state with $|\varphi|^2 \cong 1$ is superconducting.

The solutions of the equilibrium states of the Ginzburg–Landau equations are the finite energy critical points of the following action functional defined on the domain Ω in $\mathbb{R}^2$ [9]

1980 *Mathematics Subject Classification* (1985 *Revision*). 35G30.
This paper is in final form and no version of it will be submitted for publication elsewhere.

$$I_\lambda(\varphi,A) = \frac{1}{2}\int_\Omega \{(\partial_1 A_2 - \partial_2 A_1)^2 + (\partial_1\varphi_1 + A_1\varphi_2)^2 + (\partial_1\varphi_2 - A_1\varphi_1)^2 + (\partial_2\varphi_1 + A_2\varphi_2)^2 + (\partial_2\varphi_2 - A_2\varphi_1)^2 + \frac{\lambda}{4}(1 - |\varphi|^2)^2\} dx_1 dx_2 .$$

In terms of differential forms, the functional can be rewritten as

$$I_\lambda(\varphi,A) = \frac{1}{2}\int_\Omega \{|dA|^2 + |(d - iA)\varphi|^2 + \frac{\lambda}{4}(1 - |\varphi|^2)^2\} dx_1 dx_2 ,$$

where $A = A_1 dx_1 + A_2 dx_2$ is a real valued 1–form on Ω, $dA = (\partial_1 A_2 - \partial_2 A_1) dx_1 dx_2$, and $D_A\varphi = (d - iA)\varphi = (\partial_1 - iA_1)\varphi dx_1 + (\partial_2 - iA_2)\varphi dx_2$ denotes the covariant derivative of φ with respect to A.

This functional forms the basis for the mathematical results discussed in this paper. It can be easily demonstrated that the functional I_λ is invariant under the gauge transformation

(2) $$(\varphi,A) \longrightarrow (\varphi e^{i\psi}, A + \nabla\psi) ,$$

and the two physical quantities $|\varphi|^2$ and $\partial_1 A_2 - \partial_2 A_1$, the magnetic field, are also gauge invariants. These gauge invariants are critical for studying the existence of solutions of the Ginzburg–Landau equations by variational principles and the regularity of the minimizing solutions of the Ginzburg–Landau functional [9].

In a previous research paper [3] written with M. S. Berger, we studied the radially symmetric vortices, which are of the form $\varphi(r,\theta) = R(r)e^{iN\theta}$ and $A(r,\theta) = S(r)d\theta$ where N is an integer called the vortex number defined by $N = \frac{1}{2\pi}\int_\Omega dA$ [9], for the Ginzburg–Landau equations on $\mathbb{R}^2$ with arbitrary positive λ. We found that for fixed $\lambda > 0$, these equations possess a countably infinite number of distinct solutions characterized by their vortex number N. In that paper we also showed the occurrence of a linearization phenomenon for fixed N as the parameter $\lambda \longrightarrow \infty$: the nonlinear elliptic equation satisfied by the magnetic field $H_\lambda = \mathrm{curl} A_\lambda$ becomes the nonhomogeneous Helmholz equation, and $H_\lambda \longrightarrow H$ in $W_{1,p}(\mathbb{R}^2)$ for $1 \le p < 2$ where H is the solution of the nonhomogeneous Helmholz equation. Later, we extened some ideas in that paper to a more general functional, the so–called Yang–Mills–Higgs functional [5].

In the papers for the self–dual case (with fixed $\lambda = 1$), Taubes [10] found all smooth finite energy vortices of the Ginzburg–Landau equations on $\mathbb{R}^2$ without regard to symmetric considerations. However when $\lambda \neq 1$, the self–duality used by Taubes breaks down and new approaches are needed. In 1986 Bodylev [4] proved that there exist at least two gauge inequivalent solutions of the Ginzburg–Landau equations in a bounded domain in $\mathbb{R}^3$. These results on the existence of nonsymmetric solutions prompt us to consider what results in case $\lambda = 1$ can be carried over to the case $\lambda \neq 1$.

The extension to arbitrary positive λ requires a new compactness device. Cronström in 1979 [7] proved that for an arbitrary C^2 vector $A = (A_1, A_2)$, there is an explicit gauge transformation such that the new vector $\tilde{A} = (\tilde{A}_1, \tilde{A}_2)$ satisfies the Cronström gauge condition:

$$x_1 A_1(x_1,x_2) + x_2 A_2(x_1,x_2) = 0 . \tag{3}$$

If we denote A as a 1–form in Cartesian coordinates as well as polar coordinates respectively

$$A = A_1(x_1,x_2)dx_1 + A_2(x_1,x_2)dx_2 = S(r,\theta)d\theta + T(r,\theta)dr ,$$

then A_1, A_2, S and T have the following relations:

$$S = -x_2 A_1 + x_1 A_2 , \quad T = \frac{1}{r}(x_1 A_1 + x_2 A_2) .$$

Therefore, condition (3) simply means $T \equiv 0$ and $A = S(r,\theta)d\theta$. Moreover, if A satisfies both the Cronström gauge condition and the Coulomb gauge condition $\operatorname{div} A = 0$, then $A = S(r)d\theta$, i.e. A is radially symmetric, which has been studied in [3]. Utilizing the represantation $A = S(r,\theta)d\theta$, we are able to construct a Hilbert space for A in which $I_\lambda(\varphi, A)$ attains its infimum. Another remarkable feature of Cronström's gauge condition is that the vector A can be expressed explicitly in terms of $\partial_1 A_2 - \partial_2 A_1$:

$$A_1(x_1,x_2) = \int_0^1 - tx_2(\partial_1 A_2(tx_1,tx_2) - \partial_2 A_1(tx_1,tx_2))dt, \tag{4}$$

$$A_2(x_1,x_2) = \int_0^1 tx_1(\partial_1 A_2(tx_1,tx_2) - \partial_2 A_1(tx_1,tx_2))dt, \tag{5}$$

provided $x_1 A_1 + x_2 A_2 = 0$. These representations are used to obtain the regularity of the weak solutions.

2. MAIN RESULT.

By using the Cronström gauge condition and the constrained variational principle, we can prove the existence of the nonsymmetric solutions of the Ginzburg–Landau equations in a bounded convex open set Ω with C^2 boundary $\partial\Omega$ contained in $\mathbb{R}^2$. More precisely, we are able to show:

MAIN THEOREM There exist infinitely many finite energy solutions (φ,A) of (1a–1b) togather with the boundary conditions:

(1c) $$|\varphi|^2 = 1 ,$$

(1d) $$\partial_1 A_2 - \partial_2 A_1 = \text{constant} ,$$

(1e) $$(A_1 - \varphi_1\partial_1\varphi_2 + \varphi_2\partial_1\varphi_1 , A_2 - \varphi_1\partial_2\varphi_2 + \varphi_2\partial_2\varphi_1) \cdot \vec{n} = 0 ,$$

a.e. on the boundary $\partial\Omega$, where $\vec{n}$ is the normal vector to $\partial\Omega$. These solutions (φ,A) are characterized by having a nonzero total flux [1]

(1f) $$2\pi N = \int_\Omega dA = \int_\Omega (\partial_1 A_2 - \partial_2 A_1) dx_1 dx_2 ,$$

where N is an arbitrary nonzero scalar. Moreover, these solutions are not radially symmetric provided the domain Ω is not a disk in the plane.

Remark. (1d) and (1e) are the natural boundary conditions of the corresponding minimizing problem.

3. IDEA OF PROOFS.

We first prove the existence of the solutions of the minimizing problem, i.e. we show that the infimum of $I_\lambda(\varphi,A)$ over the appropriate function space is attained. Second, we show that the minimizing solutions are weak solutions and then the weak solutions are gauge equivalent to the smooth solutions of (1a–1f). Finally, we show that these solutions are nonsymmetric whenever the domain is nonradially symmetric.

We now construct the function spaces of φ and A where $I_\lambda(\varphi,A)$ attains its infimum. What are the necessary conditions on (φ,A) such that $I_\lambda(\varphi,A)$ is finite? The structure of the functional implies that if $I_\lambda(\varphi,A)$ is finite, then $*dA \in L_2(\Omega)$ and $1 - |\varphi|^2 \in L_2(\Omega)$ which implies $\varphi_1,\varphi_2 \in L_2(\Omega)$. Moreover, if A satisfies the Cronström gauge condition (3), from $A = A_1 dx_1 + A_2 dx_2 = S(r,\theta)d\theta$, we have

(6) $$A_1^2 + A_2^2 = \frac{S^2}{r^2}, \quad \text{and } *dA = \partial_1 A_2 - \partial_2 A_1 = \frac{1}{r}\partial_r S.$$

Therefore, $\frac{1}{r}\partial_r S \in L_2(\Omega)$ which implies $\frac{1}{r}S(r,\theta) \in L_2(\Omega)$ (c.f. Lemma 1), i.e. A_1 and $A_2 \in L_2(\Omega)$. Furthermore, we will claim later (Lemma 2) that we may assume $|\varphi| \le 1$ a.e. on Ω. This property with $A_1, A_2 \in L_2(\Omega)$ will give $\partial_i\varphi_j \in L_2(\Omega)$ for $i,j = 1,2$ and $\partial_i|\varphi|^2 \in L_2(\Omega)$ for $i = 1,2$. Thus, the finiteness of $I_\lambda(\varphi,A)$ and the Cronström gauge condition suggest that we might choose the following classes of functions as our solution spaces. The function space Σ_A of A is defined by

$$\Sigma_A = \{A = (A_1,A_2) \mid A_i : \Omega \longrightarrow \mathbb{R} \in L_2(\Omega),\ i = 1, 2;\ \partial_1 A_2 - \partial_2 A_1 \in L_2(\Omega),$$
$$\text{where the derivatives are in the distributional sense; and } x_1A_1 + x_2A_2 = 0 \text{ in } \Omega\}$$

with the inner product

$$(A,B)_{\Sigma_A} = \int_\Omega (\partial_1A_2-\partial_2A_1)(\partial_1B_2-\partial_2B_1), \quad \forall\, A,B \in \Sigma_A .$$

It is easy to see from (6) that Σ_A is equivalent to the space

$$\Sigma_S = \{A = S d\theta \mid S : \Omega \longrightarrow \mathbb{R},\ \tfrac{1}{r}S \in L_2(\Omega) \text{ and } \tfrac{1}{r}\partial_r S \in L_2(\Omega), \text{ where } \partial_r \text{ is}$$
$$\text{the derivative in the distributional sense}\}$$

with the inner product

$$(S_1,S_2)_{\Sigma_S} = \int_\Omega \frac{1}{r^2}(\partial_r S_1)(\partial_r S_2) .$$

The function space Σ_S has following critical properties:

<u>Lemma 1.</u>

(i) $(S_1,S_2)_{\Sigma_S} = \int_\Omega \frac{1}{r^2}(\partial_r S_1)(\partial_r S_2)$ is an inner product;

(ii) $\|\frac{1}{r}S\|_{L_2(\Omega)} \le C(\Omega)\, \|\frac{1}{r}\partial_r S\|_{L_2(\Omega)}$, for all $S \in \Sigma_S$, where $C(\Omega)$ is a constant dependent only on Ω;

(iii) Σ_S is a Hilbert space.

<u>Proof.</u> (i) We only need to check that $S = 0$ a.e. on Ω whenever $(S,S)_{\Sigma_S} = 0$.

Indeed, $(S,S)_{\Sigma_S} = 0$ implies that $\frac{1}{r}\partial_r S(r,\theta) = 0$ a.e., so that for almost all fixed θ, $\partial_r S(r,\theta) = 0$ a.e., thus, $S(r,\theta) = k(\theta)$ a.e. for almost all θ, where $k(\theta)$ is a function dependent only on θ. If $k(\theta)$ is not equal to zero almost everywhere, then, because we may assume that the origin is an interior point of Ω, there is a small $\epsilon > 0$ such that

$$\infty > \int_\Omega \frac{1}{r^2} S^2 r \, dr \, d\theta \geq \int_0^{2\pi} k^2(\theta) d\theta \int_0^\epsilon \frac{1}{r} dr = \infty ,$$

a contradiction. We have shown that $S(r,\theta) = 0$ a.e. on Ω.

(ii) From $\frac{1}{r}S \in L_2(\Omega)$ and $\frac{1}{r}\partial_r S \in L_2(\Omega)$, we have that for almost all fixed θ, $\frac{1}{\sqrt{r}} S(\cdot,\theta)$ and $\frac{1}{\sqrt{r}}\partial_r S(\cdot,\theta) \in L_2(0,r(\theta))$. Therefore $\partial_r S(\cdot,\theta) \in L_1[0,r(\theta)]$ which implies that $S(\cdot,\theta)$ is absolutely continuous on $[0,r(\theta)]$ and

$$S(r,\theta) = S(0,\theta) + \int_0^r \partial_\tau S(\tau,\theta) d\tau .$$

However $\frac{1}{\sqrt{r}}S(\cdot,\theta) \in L_2(0,r(\theta))$, thus $S(0,\theta) = 0$ and $S(r,\theta) = \int_0^r \partial_\tau S(\tau,\theta) d\tau$, for almost all θ. It follows that,

$$\begin{aligned} \|\tfrac{1}{r}S\|^2_{L_2(\Omega)} &= \int_\Omega (\tfrac{1}{r}S(r,\theta))^2 = \int_\Omega \left[\frac{1}{r}\int_0^r \partial_\tau S(\tau,\theta) d\tau\right]^2 \\ &\leq \int_\Omega \frac{1}{r^2} \left[\int_0^r \tau \, d\tau\right] \left[\int_0^{r(\theta)} \frac{(\partial_\tau S)^2}{\tau} d\tau\right] \\ &\leq C(\Omega)\|S\|^2_{\Sigma_S} . \end{aligned}$$

(iii) Let $\{S_n\}$ be a Cauchy sequence in Σ_S. By the definition of Σ_S, $\{\frac{1}{r}\partial_r S_n\}$ is a Cauchy sequence in $L_2(\Omega)$. Thus there is a function g defined on Ω, such that $\frac{1}{r}g \in L_2(\Omega)$ and $\frac{1}{r}\partial_r S_n \to \frac{1}{r}g$ in $L_2(\Omega)$. On the other hand, by (ii), $\{\frac{1}{r}S_n\}$ is a Cauchy sequence in $L_2(\Omega)$. Therefore, there is $f \in L_2(\Omega)$ to which $\frac{1}{r}S_n$ converges in $L_2(\Omega)$. We define $S = rf$. For any $\psi \in C_0^\infty(\Omega)$,

$$\begin{aligned} \int_\Omega S\partial_r(\psi r) dr d\theta &= \int_\Omega \frac{1}{r} S \partial_r(\psi r) r dr d\theta = \lim_{n\to\infty} \int_\Omega \frac{1}{r} S_n \partial_r(\psi r) r dr d\theta \\ &= -\lim_{n\to\infty} \int_\Omega \left[\frac{1}{r}\partial_r S_n\right](r\psi) r dr d\theta = -\int_\Omega g\psi r dr d\theta , \end{aligned}$$

this shows $\partial_r S = g$ in the distributional sense. We have proven that there exists a function $S \in \Sigma_S$ to which S_n converges in Σ_S.

Because of the isomorphism of Σ_A and Σ_S, we obtain the analogous properties for Σ_A immediately:

(i) $(,)_{\Sigma_A}$ is an inner product;

(ii) $\|A_1\|^2_{L_2(\Omega)} + \|A_2\|^2_{L_2(\Omega)} \le C(\Omega)\,\|dA\|^2_{L_2(\Omega)}$ for all $A \in \Sigma_A$;

(iii) Σ_A is a Hilbert space.

For function φ, according to the boundary condition $|\varphi|\Big|_{\partial\Omega} = 1$ and the necessary conditions for finiteness of $I_\lambda(\varphi,A)$, we define

$$\Sigma_\varphi = \{\varphi = \varphi_1 + i\varphi_2,\ \varphi_i : \Omega \longrightarrow \mathbb{R} \in W_{1,2}(\Omega),\ i = 1,2,\ \text{and } 1 - |\varphi|^2 \in \mathring{W}_{1,2}(\Omega)\}.$$

The function space Σ_φ has the following topological properties:

Suppose that $\{\varphi^k\}$ is a sequence in Σ_φ such that $\{\varphi^k_j\}$, $j = 1, 2$, are bounded in $W_{1,2}(\Omega)$ and $\{1 - |\varphi^k|^2\}$ is bounded in $\mathring{W}_{1,2}(\Omega)$. Then $\{\varphi^k\}$ has a subsequence still denoted by $\{\varphi^k\}$ such that

$$\varphi^k_j \longrightarrow \varphi_j \quad \text{weakly in } W_{1,2}(\Omega),\ j = 1,2,$$
$$1 - |\varphi^k|^2 \longrightarrow 1 - |\varphi|^2 \quad \text{weakly in } \mathring{W}_{1,2}(\Omega),$$

where $\varphi = \varphi_1 + i\varphi_2 \in \Sigma_\varphi$.

We notice that if $\varphi \equiv 1$ and $A \equiv 0$, then $I_\lambda(\varphi,A) = 0$. Since $I_\lambda(\varphi,A) \ge 0$ for all (φ,A), $(1,0)$ is a trivial minimizing solution of $I_\lambda(\varphi,A)$. In order to avoid the trivial solution, we utilize the total flux $2\pi N = \int_\Omega dA$ as a constraint to characterize solutions, where N is an arbitrary nonzero constant.

The result on the existence of the minimizing solutions is stated as follows.

<u>Theorem 1.</u> The infimum of $I_\lambda(\varphi,A)$ over $\Sigma_\varphi \times \Sigma_A$ with the constaint $\frac{1}{2\pi}\int_\Omega dA = N \ne 0$ is attained. Moreover, the infimum is positive.

In order to obtain the constrained minimizing solutions, besides the topological properties of Σ_A and Σ_φ discussed above, we need the boundedness for $\{\varphi^k\}$ where (φ^k, A^k) is any minimizing sequence of $I_\lambda(\varphi,A)$ over $\Sigma_\varphi \times \Sigma_A$. The maximum principle implies that if (φ,A) is a smooth solution of (1a–1c), then $|\varphi| \leq 1$ on Ω [9]. What structure of the Ginzburg–Landau functional implies this property? We found that for $|\varphi(x)| > 0$, $|d\varphi - iA\varphi|^2$ can be reformulated as

$$(7) \qquad |d\varphi - iA\varphi|^2 = \frac{1}{4|\varphi|^2}|\nabla|\varphi|^2|^2 + |\varphi|^2\Big\{\Big[\frac{1}{|\varphi|^2}(\varphi_1\partial_1\varphi_2 - \varphi_2\partial_1\varphi_1) - A_1\Big]^2 + \Big[\frac{1}{|\varphi|^2}(\varphi_1\partial_2\varphi_2 - \varphi_2\partial_2\varphi_1) - A_2\Big]^2\Big\}.$$

By virtue of formula (7) and the techniques used in the Sobolev space, we can prove

<u>Lemma 2.</u> For any $\varphi \in \Sigma_\varphi$, we define the modified function as

$$\tilde{\varphi} = \begin{cases} \varphi & \text{if } |\varphi| \leq 1, \\ \dfrac{\varphi}{|\varphi|} & \text{if } |\varphi| > 1. \end{cases}$$

Then, $\tilde{\varphi}$ has the following important properties: $|\tilde{\varphi}| \leq 1$ on Ω, $\tilde{\varphi} \in \Sigma_\varphi$, and $I_\lambda(\tilde{\varphi},A) \leq I_\lambda(\varphi,A)$, where A is an arbitrary element in Σ_A.

<u>Proof.</u> Since $|\tilde{\varphi}|^2 = \min\{1,|\varphi|^2\} = \frac{1}{2}\Big\{1 + |\varphi|^2 - \big|1-|\varphi|^2\big|\Big\}$, we have $1 - |\tilde{\varphi}|^2 = \frac{1}{2}\Big[1 - |\varphi|^2 + \big|1-|\varphi|^2\big|\Big]$. Since $1 - |\varphi|^2 \in \mathring{W}_{1,2}(\Omega)$ implies $\big|1-|\varphi|^2\big| \in \mathring{W}_{1,2}(\Omega)$, therefore, $1 - |\tilde{\varphi}|^2 \in \mathring{W}_{1,2}(\Omega)$. To prove $\tilde{\varphi}_1,\tilde{\varphi}_2 \in W_{1,2}(\Omega)$, we consider $\tilde{\varphi}_i$ as a product $\varphi_i\psi$ where ψ is the composition $f\circ|\varphi|$, f is the piecewise smooth function from $[0,\infty)$ to $\mathbb{R}$ defined by

$$f(x) = \begin{cases} 1 & x \in [0,1] \\ \dfrac{1}{x} & x \in (1,\infty). \end{cases}$$

Since $\varphi_1,\varphi_2 \in W_{1,2}(\Omega)$, we obtain

$$\partial_j|\varphi| = \begin{cases} \dfrac{\varphi_1\partial_j\varphi_1 + \varphi_2\partial_j\varphi_2}{|\varphi|} & |\varphi| > 0 \\ 0 & |\varphi| = 0, \end{cases}$$

hence $|\partial_j|\varphi|| \leq |\partial_j\varphi_1| + |\partial_j\varphi_2|$ and $|\varphi| \in W_{1,2}(\Omega)$. Thus the Chain Rule implies

$$\partial_j\psi = \partial_j(f\circ|\varphi|) = f'(|\varphi|)\partial_j|\varphi| = \begin{cases} 0 & |\varphi| \le 1 , \\ -\dfrac{\partial_j|\varphi|}{|\varphi|^2} & |\varphi| > 1 , \end{cases}$$

so that $\partial_j\psi \in L_2(\Omega)$. Clearly, $\varphi_i\psi = \tilde{\varphi}_i \in L_2(\Omega)$, and

$$\varphi_i\partial_j\psi + \psi\partial_j\varphi_i = \begin{cases} \partial_j\varphi_i & |\varphi| \le 1 \\ -\dfrac{\varphi_i\partial_j|\varphi|}{|\varphi|^2} + \dfrac{\partial_j\varphi_i}{|\varphi|} & |\varphi| > 1 \end{cases} \in L_2(\Omega),$$

therefore, applying the product rule we have

$$\partial_j\tilde{\varphi}_i = \partial_j(\varphi_i\psi) = \varphi_i\partial_j\psi + \psi\partial_j\varphi_i \in L_2(\Omega) .$$

We have proven $\tilde{\varphi}_i \in W_{1,2}(\Omega)$, $i = 1,2$.

In order to prove that $I_\lambda(\tilde{\varphi},A) \le I_\lambda(\varphi,A)$, it is sufficient to show that $\int_{|\varphi|>1} |(d-iA)\tilde{\varphi}|^2 \le \int_{|\varphi|>1} |(d-iA)\varphi|^2$ because the other terms are obvious. Indeed, since on the set $\{x \in \Omega, |\varphi(x)|>1\}$, $\nabla|\tilde{\varphi}|^2 \equiv 0$ and

$$\frac{1}{|\tilde{\varphi}|^2}(\tilde{\varphi}_1\partial_i\tilde{\varphi}_2 - \tilde{\varphi}_2\partial_i\tilde{\varphi}_1) = \frac{1}{|\varphi|^2}(\varphi_1\partial_i\varphi_2 - \varphi_2\partial_i\varphi_1), \qquad i = 1,2 ,$$

formula (7) implies the inequality.

Now we prove Theorem 1.

<u>Proof of Theorem1.</u> Let $\Sigma_N = \{ (\varphi,A),\ \varphi \in \Sigma_\varphi,\ A \in \Sigma_A$, and A satisfies the constraint $N = \frac{1}{2\pi}\int_\Omega dA$, where $N \ne 0\}$. Suppose that (φ^k,A^k) is a minimizing sequence of $I_\lambda(\varphi,A)$ over Σ_N , that is $\lim_{k\to\infty} I_\lambda(\varphi^k,A^k) = \inf_{(\varphi,A)\in\Sigma_N} I_\lambda(\varphi,A)$. Without loss of generality, we may assume that $|\varphi^k| \le 1$ a.e. on Ω and $I_\lambda(\varphi^k,A^k) \le M$ for all k, where M is a positive number. Since

$$I_\lambda(\varphi,A) = \tfrac{1}{2}[\ \|dA\|^2_{L_2(\Omega)} + \|(d-iA)\varphi\|^2_{L_2(\Omega)} + \tfrac{\lambda}{4}\|1-|\varphi|^2\|^2_{L_2(\Omega)}\ ],$$

we have $\{A^k\}$ is bounded in the Hilbert space Σ_A . Since $|\varphi| \le 1$ and $\|A_1\|^2_{L_2(\Omega)} + \|A_2\|^2_{L_2(\Omega)} \le C\|dA\|^2_{L_2(\Omega)} \le 2CM$, we have $\{A^k_i\varphi^k_j\}$ is bounded in $L_2(\Omega)$, it follows that $\{\varphi^k_j\}$ is bounded in $W_{1,2}(\Omega)$. Therefore, there exists both a subsequence still denoted by $\{\varphi^k, A^k\}$, and $(\varphi, A) \in \Sigma_\varphi \times \Sigma_A$, such that

$$\begin{aligned}
&A^k \longrightarrow A \quad \text{weakly in } \Sigma_A\,, \\
&\varphi_j^k \longrightarrow \varphi_j \quad \text{in } L_p(\Omega), \quad \text{where } 1 \le p < \infty\,, \\
&\varphi_j^k \longrightarrow \varphi_j \quad \text{a.e. on } \Omega\,, \\
&\nabla\varphi_j^k \longrightarrow \nabla\varphi_j \quad \text{weakly in } L_2(\Omega)\,, \\
&A_i^k\varphi_j^k \longrightarrow A_i\varphi_j \quad \text{weakly in } L_2(\Omega)\,.
\end{aligned}$$

Using the properties of weakly convergent sequences and Fatou's Lemma, we obtain that

$$I_\lambda(\varphi,A) \le \varliminf_{k\to\infty} I_\lambda(\varphi^k,A^k) = \inf_{(\varphi,A)\in\Sigma_N} I_\lambda(\varphi,A).$$

Since $A^k \longrightarrow A$ weakly in Σ_A, by the definition of weak convergence, we have

$$\frac{1}{2\pi}\int_\Omega (\partial_1 A_2 - \partial_2 A_1) = \lim_{k\to\infty} \frac{1}{2\pi}\int_\Omega (\partial_1 A_2^k - \partial_2 A_1^k) = \lim_{k\to\infty} N = N\,.$$

Thus, $(\varphi, A) \in \Sigma_N$ which implies $I_\lambda(\varphi,A) = \inf_{(\varphi,A)\in\Sigma_N} I_\lambda(\varphi,A)$.

To show $I_\lambda(\varphi,A) > 0$, we assume $I_\lambda(\varphi,A) = 0$ to obtain a contradiction. From $I_\lambda(\varphi,A) = 0$, we have $\int_\Omega (\partial_1 A_2 - \partial_2 A_1)^2 = 0$, so that $\int_\Omega (\partial_1 A_2 - \partial_2 A_1) = 0$. This contradicts $\int_\Omega (\partial_1 A_2 - \partial_2 A_1) = 2\pi N$ where N is a nonzero constant.

Next, we give the regularity result. We first define the weak solution for our problem.

Definition. A weak solution of the Ginzburg–Landau equations (1a–1b) is a pair $(\varphi, A) \in \Sigma_\varphi \times \Sigma_A$ such that for all $\xi_1, \xi_2, \eta_1, \eta_2 \in C_0^\infty(\Omega)$, the Gateaux derivative of I_λ at (φ, A) in the direction $(\xi_1 + i\xi_2\,,\ \eta_1 dx_1 + \eta_2 dx_2)$ equals zero, or equivalently,

$$\begin{aligned}
0 = \int_\Omega &\partial_1\xi_1(\partial_1\varphi_1 + A_1\varphi_2) + \partial_2\xi_1(\partial_2\varphi_1 + A_2\varphi_2) \\
&+ \xi_1\Big[(A_1^2 + A_2^2)\varphi_1 + \tfrac{\lambda}{2}(|\varphi|^2 - 1)\varphi_1 - A_1\partial_1\varphi_2 - A_2\partial_2\varphi_2\Big]\,,
\end{aligned}$$

$$0=\int_\Omega \partial_1\xi_2(\partial_1\varphi_2-A_1\varphi_1)+\partial_2\xi_2(\partial_2\varphi_2-A_2\varphi_1)$$
$$+\xi_2\Big[(A_1^2+A_2^2)\varphi_2+\tfrac{\lambda}{2}(|\varphi|^2-1)\varphi_2+A_1\partial_1\varphi_1+A_2\partial_2\varphi_1\Big],$$

$$0=\int_\Omega \partial_2\eta_1(\partial_2A_1-\partial_1A_2)+\eta_1(A_1|\varphi|^2+\varphi_2\partial_1\varphi_1-\varphi_1\partial_1\varphi_2),$$

$$0=\int_\Omega -\partial_1\eta_2(\partial_2A_1-\partial_1A_2)+\eta_2(A_2|\varphi|^2+\varphi_2\partial_2\varphi_1-\varphi_1\partial_2\varphi_2).$$

Theorem 2. The constrained minimizing solution (φ,A) of I_λ over Σ_N obtained in Theorem 1 is a weak solution.

Proof. The result is obtained by showing that for any $\xi=\xi_1+i\xi_2$ and $\eta=(\eta_1,\eta_2)$, where ξ_1, ξ_2, η_1 and $\eta_2\in C_o^\infty(\Omega)$, we have

$$0=\lim_{t\to o}\frac{I_\lambda(\varphi+t\xi,A)-I_\lambda(\varphi,A)}{t}$$
$$=\int_\Omega (\partial_1\varphi_1+A_1\varphi_2)(\partial_1\xi_1+A_1\xi_2)+(\partial_2\varphi_1+A_2\varphi_2)(\partial_2\xi_1+A_2\xi_2)$$
$$+(\partial_1\varphi_2-A_1\varphi_1)(\partial_1\xi_2-A_1\xi_1)+(\partial_2\varphi_2-A_2\varphi_1)(\partial_2\xi_2-A_2\xi_1)$$
$$-\frac{\lambda}{2}(1-|\varphi|^2)(\varphi_1\xi_1+\varphi_2\xi_2),$$

and

$$0=\lim_{t\to 0}\frac{I_\lambda(\varphi,A+t\eta)-I_\lambda(\varphi,A)}{t}$$
$$=\int_\Omega (\partial_2A_1-\partial_1A_2)(\partial_2\eta_1-\partial_1\eta_2)$$
$$+\eta_1(A_1|\varphi|^2+\varphi_2\partial_1\varphi_1-\varphi_1\partial_1\varphi_2)+\eta_2(A_2|\varphi|^2+\varphi_2 2_2\varphi_1-\varphi_1\partial_2\varphi_2).$$

The first limit follows from straightforward computation as well as the inequality $I_\lambda(\varphi+t\xi,A)\geq \inf_{(\varphi,A)\in\Sigma_N} I_\lambda(\varphi,A)=I_\lambda(\varphi,A)$ because $(\varphi+t\xi,A)\in\Sigma_N$ for all $t\in\mathbb{R}$. In order to obtain the second limit, the key step is to show that $I_\lambda(\varphi,A+t\eta)\geq I_\lambda(\varphi,A)$ for all t. This can be achieved by virtue of Cronström gauge condition and the property that $I_\lambda(\varphi,A)$ is a gauge invariant under the gauge transformation (2). Indeed, since $\eta_1,\eta_2\in C_o^\infty(\Omega)$, according to the result given by Cronström [7], there is smooth function Ψ

on $\overline{\Omega}$, such that the new vector function $\tilde{\eta} = (\tilde{\eta}_1, \tilde{\eta}_2) = (\eta_1 + \partial_1 \Psi, \eta_2 + \partial_2 \Psi)$ satisfies the Cronström gauge condition (3)

$$x_1 \tilde{\eta}_1 + x_2 \tilde{\eta}_2 = 0 .$$

Thus, $t\tilde{\eta}$ satisfies (3) for all $t \in \mathbb{R}$, hence $A + t\tilde{\eta} \in \Sigma_A$. Let $\tilde{\varphi}^t = \varphi e^{it\Psi}$. Since $\Psi \in C^\infty(\overline{\Omega})$ and $|\tilde{\varphi}^t| = |\varphi|$, we have $\tilde{\varphi}^t \in \Sigma_\varphi$. Since $\int_\Omega d(A + t\tilde{\eta}) = \int_\Omega (\partial_1 A_2 - \partial_2 A_1) + t(\partial_1 \eta_2 - \partial_2 \eta_1)$, and $\eta_1, \eta_2 \in C_0^\infty(\Omega)$ which implies $\int_\Omega (\partial_1 \eta_2 - \partial_2 \eta_1) = 0$, we have $\int_\Omega d(A + t\tilde{\eta}) = 2\pi N$. Thus, $(\tilde{\varphi}^t, A + t\tilde{\eta}) \in \Sigma_N$. Therefore,

$$I_\lambda(\tilde{\varphi}^t, A + t\tilde{\eta}) \geq \inf_{(\varphi,A)\in\Sigma_N} I_\lambda(\varphi,A) = I_\lambda(\varphi,A) .$$

Since the functional $I_\lambda(\varphi,A)$ is gauge invariant, that is

$$I_\lambda(\varphi,A + t\eta) = I_\lambda(\varphi e^{it\Psi}, A + t\eta + \nabla t\Psi) = I_\lambda(\tilde{\varphi}^t, A + t\tilde{\eta})$$

for all $t \in \mathbb{R}$, we obtain $I_\lambda(\varphi,A + t\eta) \geq I_\lambda(\varphi,A)$.

Remark. In general, using the constrained (or isoperimetric) variational principle, we will obtain extra terms (corresponding to the constraints) in the equations satisfied by the weak solutions compared with the original equations. However, in the problem considered here, we don't have the extra term because of the property that for all $\eta_1, \eta_2 \in C_0^\infty(\Omega)$, $\int_\Omega \partial_1 \eta_2 - \partial_2 \eta_1 = 0$. In such a case the constraint is called a natural constraint. The variational principle with natural constraints has been proven to be very useful in many applications [2].

Based on the regularity result given by Jaffe and Taubes [9] and the representations of A_1 and A_2 (formulas (4–5)), we can prove that

Theorem 3. Suppose that (φ,A) is a minimizing solution proven to exist in Theorem 1. Then (φ,A) is guage equivalent to a pair of smooth functions $(\tilde{\varphi},\tilde{A})$ (i.e. $\tilde{\varphi}_1$, $\tilde{\varphi}_2$, $\tilde{A}_1$, $\tilde{A}_2 \in C^\infty(\Omega)$) which satisfies (1a–1f) in Ω.

The proof of Theorem 3 is tedious, it can be found in [6].

Finally, combining the maximum principle with the representation of $|(d-iA)\varphi|^2$ (formula (7)) and the theorem related to the winding number, we obtain

Theorem 4. If the domain Ω is not a disk, then the solution $(\tilde{\varphi},\tilde{A})$ found in Theorem 3 is not radially symmetric.

Proof. ($\sim$ is suppressed in the proof) It is sufficient to prove $1-|\varphi|^2>0$ in Ω. Suppose that at $x_o\in\Omega$, $1-|\varphi(x_o)|^2=0$. Since (φ,A) is a smooth solution of (1a–1b), it can be shown [9] that (φ,A) satisfies

$$\Delta\tfrac{1}{2}(1-|\varphi|^2)-\tfrac{\lambda}{2}|\varphi|^2(1-|\varphi|^2)=-|(d-iA)\varphi|^2 \text{ in } \Omega .$$

Since $1-|\varphi|^2\geq 0$ in Ω, the Maximum Principle [9] implies $1-|\varphi|^2\equiv 0$ in Ω_1 where Ω_1 is an open subset of Ω such that $x_o\in\Omega_1$ and $\overline{\Omega}_1\subset\Omega$. Therefore, $1-|\varphi|^2\equiv 0$ in Ω, which implies $|(d-iA)\varphi|^2\equiv 0$ in Ω. Thus, by formula (7), we obtain

$$A_1=\varphi_1\partial_1\varphi_2-\varphi_2\partial_1\varphi_1 , \quad A_2=\varphi_1\partial_2\varphi_2-\varphi_2\partial_2\varphi_1 .$$

Let $\{\Omega_k\}$ be a sequence of subset of Ω such that $\partial\Omega_k$ are smooth and Ω_k approaches Ω as k approaches infinity. Then, the smoothness of φ and A imply

$$\begin{aligned}2\pi N&=\int_\Omega dA=\lim_{k\to\infty}\int_{\Omega_k} dA=\lim_{k\to\infty}\int_{\partial\Omega_k} A\\&=\lim_{k\to\infty}\int_{\partial\Omega_k}(\varphi_1\partial_1\varphi_2-\varphi_2\partial_1\varphi_1)dx_1+((\varphi_1\partial_2\varphi_2-\varphi_2\partial_2\varphi_1)dx_2 .\end{aligned}$$

On the other hand, since φ_1 and $\varphi_2\in C^\infty(\Omega)$ and $|\varphi|^2\equiv 1$ on Ω, we obtain [9]

$$\int_{\partial\Omega_k}(\varphi_1\partial_1\varphi_2-\varphi_2\partial_1\varphi_1)dx_1+(\varphi_1\partial_2\varphi_2-\varphi_2\partial_2\varphi_1)dx_2=0 ,$$

for all k. This contradiction concludes the proof.

BIBLIOGRAPHY

[1] A. A. Abrikosov, "On the magnetic properties of superconductors of the second group," Sov. Phys. JETP 5(1957), 1174–1182.

[2] M. S. Berger, Nonlinearity and functional analysis, Academic Press, New York, 1975.

[3] M. S. Berger, Y. Y. Chen, "Symmetric vortices for theGinzburg–Landau Equations of Superconductivity and the nonlinear desingularization phenomenon," J. Funct. Anal., Vol 82, No. 2(1988), 259–295.

[4] N. A. Bodylev, "Topological index of extremals of multidimensional variational problems," Functions Analysis and its Applications, Vol 20, No. 2(1986), 89–93.

[5] Y. Y. Chen, "The nonlinear desingularization phenomenon in the Abelian Higgs model for isotropic solutions," J. Math. Phys., Vol. 29, No. 8(1988), 1896–1899.

[6] Y. Y. Chen, "Nonsymmetric vortices for the Ginzburg–Landau Equations on the bounded domain," (to be published in J. Math. Phys.).

[7] C. Cronström, "A simple and complete Lorentz–covariant gauge condition," Phys. Lett 90B, (1980), 267–269.

[8] V. L. Ginzburg, L. D. Landau, "On the theory of superconductivity," Zh. Eksper. Teoret. Fiz. 20(1950), 1064–1073.

[9] A. Jaffe, C. Taubes, Vortices and monopoles. Birkhäuser, Basel, (1980).

[10] C. Taubes, "Arbitrary n–vortex solutions to the first order Ginzburg–Landau equations," Comm. Math. Phys. 72(1980), 277–292. "On the equivalence of the first and second order equations for gauge theories," Comm. Math. Phys. 75(1980), 207–227.

DEPARTMENT OF MATHEMATICS
CALIFORNIA STATE UNIVERSITY
5500 UNIVERCITY PARKWAY
SAN BERNARDINO, CA 92407

Contemporary Mathematics
Volume **108**, 1990

NONLINEAR STABILITY AND BIFURCATION IN HAMILTONIAN SYSTEMS WITH SYMMETRY

Andrew Szeri & Philip Holmes

ABSTRACT. We consider infinite dimensional Hamiltonian systems with symmetry. We assume that components of the equilibrium fields are related by what we call a structure function, which arises naturally in applications. Nonlinear stability criteria are established via the energy-Casimir method of Arnold. Under certain restrictions, we demonstrate that stability is shared by distinct equilibria within the same *family* (equilibria which have the same structure function). When bifurcation occurs, the stability of secondary branches may be inferred from that of the primary branch for perhaps large ranges of the parameter, regardless of the type of bifurcation. As an example, we consider bifurcation of an axisymmetric swirling flow of an ideal fluid.

1. INTRODUCTION. The energy-Casimir method of Arnold and others (Arnold (1965,1969), Holm et al. (1985)) provides a framework for generating a Lyapunov-type stability argument for equilibria of infinite dimensional Hamiltonian dynamical systems. The method is based on constructing a conserved quantity which has a convex critical point at an equilibrium of the dynamical system. This

---

The final version of this paper will be published elsewhere.
1980 Mathematics Subject Classification (1985 Revision). 76E30, 35B35, 35B32.

provides a norm in terms of which one can show that solutions started near the equilibrium remain near it for all time, and hence that the equilibrium is nonlinearly stable.

The conserved quantity of Arnold's method is normally constructed from the Hamiltonian and from other constants of the motion which correspond, via Noether's Theorem, to symmetries of the dynamical system. The method has a better chance of success if there are many such constants of the motion, and (especially) if they contain undetermined parameters or functions. See Holm et al. (1985) for numerous examples, as well as Abarbanel et al. (1986), Lewis (1989) and Szeri & Holmes (1988), among others.

The energy-Casimir method yields sufficient conditions for nonlinear stability, i.e. satisfaction of the criteria guarantees nonlinear stability, but stability may also obtain if the criteria fail. Because the criteria are sufficient, but not necessary, the relationship between nonlinear stability and bifurcation of the underlying dynamical system is an awkward one to establish. This difficulty is compounded by symmetries of the Hamiltonian system. It is well known that normally improbable and even bizarre types of bifurcation may be generic in the presence of symmetry; see Golubitsky & Schaeffer (1985, 1988) for various interesting examples of this phenomenon.

In this paper, we address the issue of the relationship between nonlinear stability criteria obtained by the energy-Casimir method and bifurcations to other relative equilibria of the underlying dynamical system. We focus on the case of an infinite dimensional noncanonical Hamiltonian system. Such systems are common in fluid dynamics, plasma dynamics, etc. We endow our Hamiltonian system with an additional piece of structure, which arises quite naturally for examples in which the equilibrium equations for the dynamical system are identically satisfied when certain functional relationships exist among the equilibrium fields. For instance in two dimensional incompressible fluid flow, functional dependence of the point values of the vorticity field on the point values of the stream function field implies that the stream function is that of an equilibrium flow.

In section two of this paper, we analyze a class of Hamiltonian systems with symmetry which possess functional dependence among their steady solutions. We generate sufficient conditions for formal and nonlinear stability using the energy-Casimir method. We give precise definitions of these terms below. In section three, we show that stability results from one equilibrium extend to other equilibria which share the same structure function and range of the dependent variable, under certain restrictions on the Hamiltonian system (Theorem I). Next we apply these ideas in section four to the analysis of bifurcation problems. We examine bifurcations from stable branches of equilibria to new branches of equilibria which satisfy the same functional relationships among the equilibrium fields. We prove in Theorem II that in a large class of problems, the stability of bifurcating branches may be inferred from that of the primary branch, without regard to the type of bifurcation. In section five, we consider the example of axisymmetric swirling flows of an ideal fluid, the Hamiltonian structure of which was described by Szeri & Holmes (1988), hereafter referred to as S&H. We show how the abstract arguments of section two are applied, and we carry the example through to a bifurcation problem which illustrates the use of Theorem II. Unfortunately, the example is physically unrealistic, because a drastic high wavenumber cut off must be applied to obtain stability criteria. Finally, we give our conclusions in section six. A more detailed treatment of this topic together with more physical examples will be given in Szeri (1989).

2. STABILITY ANALYSIS OF THE HAMILTONIAN SYSTEM. We consider infinite dimensional noncanonical Hamiltonian systems which are symmetric with respect to the action of a Lie group. We will work in the reduced phase space, i.e. the dependent variables belong to the dual of the Lie algebra of the symmetry group $\mathfrak{g}$, so that $\mu \in \mathfrak{g}^*$. We view $\mu(x,t)$ as a function over some (physical) domain $D\times\mathcal{R}$. The

Hamiltonian is a function on the dual of the Lie algebra $H:\mathfrak{g}^*\to\mathbb{R}$; corresponding to the symmetry there is a conserved quantity (or quantities) $C:\mathfrak{g}^*\to\mathbb{R}$. We remark that in practice, C may involve undetermined functions or constants as well as the dependent variable. In fact, for the purposes of establishing nonlinear stability, it is enough that C be a constant of the motion ($dC/dt=0$) when evaluated along orbits of the hamiltonian vector field at hand; see S&H. Nonetheless, we shall still refer to the technique as the energy-Casimir method. The evolution equations for μ have the form

$$\frac{d\mu}{dt} = X_H(\mu), \tag{2.1}$$

where X_H is generally a nonlinear operator.

Finally, we assume that there is a (linear) projection operator P and a function $\mathcal{F}$ which satisfy the following relation:

$$(\mathrm{id} - P)\mu_e = \mathcal{F}(P\mu_e,\lambda) \Rightarrow X_H(\mu_e) = 0, \tag{2.2}$$

where $\mu_e(x;\lambda)$ is an equilibrium value of μ, which depends on x and possibly on the parameter(s) λ. In the examples known to us so far, P is simply a projection which splits the (generally vector-valued) field μ into its different component fields. Thus $\mathcal{F}$, which we shall call the *structure function* corresponding to μ_e and P, emerges as a functional relationship between different components of the equilibrium field μ_e. Equation (2.2) states that this functional relationship implies that μ_e is an equilibrium. Note that such functional dependence is not necessary for existence of an equilibrium. We define $\rho=P\mu$, $\sigma=(\mathrm{id}-P)\mu$, $\delta\rho=P\delta\mu$ and $\delta\sigma=(\mathrm{id}-P)\delta\mu$ for use below.

2.1 THE ARNOLD FUNCTION. Following the energy-Casimir method, we form the candidate Arnold function from the Hamiltonian and other constant(s) of the motion by setting $H_C(\mu)=H(\mu)+C(\mu)$. We require that the Arnold function have a critical point at an equilibrium of the dynamical system. Taking the first variation of H_C with respect to variations $\delta\mu\in\mathfrak{g}^*$, we obtain

$$\delta H_C(\mu) = DH(\mu) \bullet \delta\mu + DC(\mu) \bullet \delta\mu, \tag{2.3}$$

where the derivatives are defined by

$$\frac{d}{d\varepsilon}F(\mu + \varepsilon\delta\mu)|_{\varepsilon=0} = DF(\mu) \bullet \delta\mu. \tag{2.4}$$

We require that the first variation evaluated at an equilibrium μ_e vanish, which yields

$$0 = DH(\mu_e) \bullet \delta\mu + DC(\mu_e) \bullet \delta\mu. \tag{2.5}$$

Next we rewrite equation (2.5) as

$$DC(\mu_e) \bullet \delta\mu = -DH(\mu_e) \bullet \delta\mu \tag{2.6}$$

and use the projection P and the function $\mathcal{F}$ of equation (2.2) to obtain

$$DC(\rho_e + \mathcal{F}(\rho_e,\lambda)) \bullet \delta\mu = -DH(\rho_e + \mathcal{F}(\rho_e,\lambda)) \bullet \delta\mu. \tag{2.7}$$

Usually, equation (2.6) serves to fix some or all of the undetermined functions or constants in the general conserved quantity C. The specialized version of C, defined via (2.7), we shall denote with a hat. Next, we *choose* to extend this definition globally away from μ_e. This yields

$$D\hat{C}(\rho) \bullet \delta\mu = -DH(\rho + \mathcal{F}(\rho,\lambda)) \bullet \delta\mu, \tag{2.8}$$

where

$$\hat{C}(\rho) \equiv C(\rho + \mathcal{F}(\rho,\lambda)) \tag{2.9}$$

is the specialization of C which yields $\delta H_C(\mu_e)=0$. Note that both the structure function and the form of the Hamiltonian have influenced the Arnold function through the requirement that there be a critical point at equilibrium. It is this structure which we are going to exploit.

2.2 SECOND VARIATION AND FORMAL STABILITY CRITERIA.

Reconstructing the first variation away from equilibrium, we have

$$\begin{aligned}\delta H_{\hat{C}}(\mu) &= DH(\mu) \bullet \delta\mu + D\hat{C}(\mu) \bullet \delta\mu \\ &= DH(\mu) \bullet \delta\mu - DH(\rho + \mathcal{F}(\rho,\lambda)) \bullet \delta\mu.\end{aligned} \tag{2.10}$$

As required, this first variation vanishes at an equilibrium μ_e, as is easily shown using equation (2.2).

The second variation is readily computed to be

$$\begin{aligned}\delta^2 H_{\hat{C}}(\mu) &= D^2H(\mu) \bullet (\delta\mu,\delta\mu) \\ &- D^2H(\rho + \mathcal{F}(\rho,\lambda)) \bullet (\delta\rho + D\mathcal{F}(\rho,\lambda) \bullet \delta\rho,\delta\mu),\end{aligned} \tag{2.11}$$

where $D\mathcal{F}(\rho,\lambda)$ is the derivative with respect to the first argument. Evaluating at the equilibrium μ_e, we obtain

$$\delta^2 H_{\hat{C}}(\mu_e) = D^2H(\mu_e) \bullet (\delta\mu,\delta\mu) - D^2H(\mu_e) \bullet (\delta\rho + D\mathcal{F}(\rho_e,\lambda) \bullet \delta\rho,\delta\mu). \tag{2.12}$$

To prove formal stability of the equilibrium in question, we must establish the positive (or negative) definiteness of the quadratic form of equation (2.12). Thus we must prove that

$$D^2H(\mu_e) - D^2H(\mu_e) \bullet (P + D\mathcal{F}(P\mu_e,\lambda) \bullet P) \tag{2.13}$$

is positive (or negative) definite. Note that definiteness of (2.13) depends on the derivatives of the Hamiltonian and the structure function and on the projection P.

We remark that formal stability of μ_e is equivalent to Lyapunov stability of the evolution equations linearized about μ_e; thus formal stability implies linear stability. The stronger condition of nonlinear stability requires the construction of a (local) norm near μ_e and will be considered below. Nonlinear stability holds if for every $\varepsilon>0$ there is a $\delta>0$ such that if $||\mu(0)-\mu_e||<\delta$, then $||\mu(t)-\mu_e||<\varepsilon$ for all $t>0$, where $||.||$ denotes a suitable norm. Note that formal stability does not necessarily imply nonlinear stability (nor does nonlinear stability imply formal stability).

2.3 SUFFICIENT CONDITIONS FOR NONLINEAR STABILITY. We now derive sufficient conditions for the convexity of the Arnold function near an equilibrium. See Holm, et al. (1985) for a careful discussion of this part of the analysis for general systems. To begin, we consider a lower bound for the Hamiltonian in the neighborhood of the equilibrium: that is a quadratic form $Q_1(\Delta\mu)$ such that

$$Q_1(\Delta\mu) \le D^2H(\mu) \bullet (\Delta\mu,\Delta\mu) \tag{2.14}$$

for all $\mu\in\mathrm{range}\{\mu_e(x)\}$. Here $\Delta\mu$ is a small but finite perturbation. Also, we bound the constant(s) of the motion from below by $Q_2(\Delta\mu)$:

(2.15) $$Q_2(\Delta\mu) \le D^2\hat{C}(\mu) \bullet (\Delta\mu,\Delta\mu)$$

or

(2.16) $$Q_2(\Delta\mu) \le -D^2H(\mu) \bullet (P + D\mathcal{F}(P\mu,\lambda) \bullet P) \bullet (\Delta\mu,\Delta\mu)$$

for all $\mu \in \text{range}\{\mu_e(x)\}$. Next, we require that

(2.17) $$\|\Delta\mu\|^2 \equiv Q_1(\Delta\mu) + Q_2(\Delta\mu) > 0$$

for all $\Delta\mu \neq 0$. Finally, we have nonlinear stability if (2.17) is true and if we have the upper bounds

(2.18) $$D^2H(\mu) \bullet (\Delta\mu, \Delta\mu) \le C_1\|\Delta\mu\|^2$$

and

(2.19) $$D^2\hat{C}(\mu) \bullet (\Delta\mu,\Delta\mu) \le C_2\|\Delta\mu\|^2$$

for all $\mu \in \text{range}\{\mu_e(x)\}$. This allows the a priori stability estimate

(2.20) $$\|\Delta\mu(t)\|^2 \le (C_1 + C_2)\|\Delta\mu(t=0)\|^2$$

for all $t>0$. Note that the inequalities which are sufficient conditions for nonlinear stability, equations (2.14,16-19), involve derivatives of the Hamiltonian and of the structure function, and the projection P all evaluated over the *range* of $\mu_e(x)$.

3. STABILITY OF RELATED EQUILIBRIA. Upon examination, the stability criteria for nonlinear (and formal) stability are found to be inequalities of the form

(3.1) $$\delta^- \le K(\mu_e(x;\lambda), x,\lambda) \le \delta^+.$$

In the special case in which *each* inequality that makes up the nonlinear stability criteria may be decomposed in the form

(3.2) $$\delta^- \le F(\mu_e(x;\lambda), \lambda) + G(x, \lambda) \le \delta^+,$$

where either F or G (but not both) may be zero, we show below that stability criteria computed for one equilibrium hold for other, related equilibria. The success of the decomposition of K into F+G depends on the form of the Hamiltonian and on the structure function. We shall call Arnold functions which lead to the decomposition of stability criteria of the form (3.1) into the form (3.2) *decomposable* Arnold functions.

3.1 STABILITY OF TWO EQUILIBRIA IN THE SAME FAMILY. In what follows, we concentrate on nonlinear rather than on formal stability, although it is not difficult to derive results parallel to the ones we give below for the case of formal stability. First we define a family of equilibria to be the collection of all equilibria having the same structure function $\mathcal{F}$ and projection P, that is

$$\text{family}(\mathcal{F},P) \equiv \{\mu \in \mathfrak{g}^* : (\text{id} - P)\mu = \mathcal{F}(P\mu,\lambda), \lambda \in \Lambda\}, \tag{3.3}$$

where Λ is the space of parameter(s). Note that the hypothesis (2.2) states that $\mu \in \text{family}(\mathcal{F},P)$ for some $\mathcal{F}$ and P implies that $X_H(\mu)=0$, i.e. μ is an equilibrium of the evolution equations. Thus a family includes all of the solutions of a given structure function equation for all values of the parameter(s).

For distinct equilibria within the same family and at the same value of the parameter, we have the following stability result:

THEOREM I: (Stability within a family)
Consider a Hamiltonian system which leads to a decomposable Arnold function. Suppose we have two equilibria belonging to the same family, that is μ_e^a, μ_e^b are both in family(F,P). Suppose further that $\rho_e^a = P\mu_e^a$ and $\rho_e^b = P\mu_e^b$ have the same range over the domain D at the same value of the parameter λ. Then if μ_e^a is nonlinearly stable by the criteria of equation (2.14,16-19), μ_e^b is also nonlinearly stable at the same value of the parameter.

We sketch the proof of this theorem as follows. By hypothesis, ρ_e^a and ρ_e^b have the same ranges in D. Because μ_e^a and μ_e^b belong to the same family, and therefore have the same structure function, the ranges of μ_e^a and μ_e^b also coincide, by equation (2.2). Thus any inequality of the form (3.2) will be satisfied by μ_e^b if it is satisfied by μ_e^a. Now in the case where the Arnold function is decomposable the stability criteria (equations (2.14,16-19)) take forms analogous to equation (3.2). Thus stability of $\mu_e{}^a$ implies stability of $\mu_e{}^b$.

A remark about applications is in order. In many examples, terms like $D^2H(\mu_e)$ may contain operators, which are normally replaced by bounds during the stability analysis through the use of a Poincaré-type inequality; see Holm et al. (1985) for examples. In this case, Theorem I would normally apply to the stability criteria after replacement of the operator terms by their bounds.

4. BIFURCATIONS OF STABLE BRANCHES OF EQUILIBRIA. We shall now pursue the ramifications of Theorem I for problems of bifurcation within the same family. In order to profit from the preceding analysis, we shall restrict our consideration to systems with decomposable Arnold functions. To be specific, we examine the case where $\mu_e{}^b=\mu_e{}^a+\nu$, and ν satisfies the bifurcation problem

(4.1) $$(\mathrm{id}-\mathrm{P})(\mu_e^a+\nu)=\mathcal{F}(\mathrm{P}(\mu_e^a+\nu),\lambda)$$

for a given equilibrium $\mu_e{}^a(x;\lambda)$, $\mathcal{F}$, and P. Thus we consider bifurcations to other solutions within the same family. Equation (4.1) may turn out to be an algebraic equation, or more generally an ordinary or partial differential equation.

We remark that equation (4.1) forms an important class of problems in applications. As an example, consider the case of two-dimensional incompressible fluid flow between parallel solid boundaries. In this problem, $\mu=\omega(x_1,x_2)$ is the scalar vorticity, related uniquely to the stream function ψ through the Laplace operator, i.e. we have $-\nabla^2\psi_e=\omega_e$. Alternately, $\psi_e=(-\nabla^2)^{-1}\omega_e$, with the appropriate boundary conditions. In this example, the projection operator is the identity and we have the structure function

$$0=\mathcal{F}(\omega_e)=\frac{dB}{d\psi}\left(\left(-\nabla^2\right)^{-1}\omega_e\right)-\omega_e.$$

Here $B(\psi)$ is the Bernoulli function of the fluid flow, defined by

$$B(\psi(x_1,x_2)) = \hat{B}(x_1,x_2) = \frac{p}{\rho} + \frac{1}{2}q^2,$$

where p is the pressure, ρ is the fluid density and q is the speed of the fluid. Bifurcations of equation (4.1) correspond in this example to the existence of multiple steady flows with the same flow rate and the same Bernoulli function. Often in fluid dynamics one treats the problem where the Bernoulli function, or $\mathcal{F}$ in general, is determined "far upstream". See Batchelor (1956), Benjamin (1962,1984), Squire (1956) and Leibovich (1970,1987) for examples of the rich variety of applications of this idea. The example of section five is also of this type.

Returning to the analysis of the general bifurcation problem, it is clear that for a given structure function $\mathcal{F}$ and projection P, we require that

$$\text{(4.2)} \qquad \underset{D}{\mathrm{range}}\left[\mu_e^a(x;\lambda)\right] = \underset{D}{\mathrm{range}}\left[\mu_e^a(x;\lambda) + v(x;\lambda)\right]$$

in order to infer the stability of μ_e^a+v from that of μ_e^a. Rewriting equation (4.1) using equation (2.2), we have

$$\text{(4.3)} \qquad (\mathrm{id}-P)(v) - \left[\mathcal{F}(P(\mu_e^a+v),\lambda) - \mathcal{F}(P(\mu_e^a),\lambda)\right] = 0.$$

In the case where (4.3) is a differential equation, the normal homogeneous boundary conditions on v are helpful in testing whether or not equation (4.2) holds, as we show below.

Now, the second variation of the Arnold function must be indefinite *at* the point of bifurcation in the "direction(s)" of bifurcation. To see this, note that at a bifurcation , the linearized form of equation (4.3) admits a nontrivial solution $\delta\mu^*$ which satisfies

$$\text{(4.4)} \qquad \left[(\mathrm{id}-P) - D\mathcal{F}(\mu_{2e},\lambda_c)\bullet P\right]\bullet\delta\mu^* = 0.$$

Equation (2.12) gives the second variation which we evaluate in the direction of the bifurcation:

$$\text{(4.5)} \qquad D^2H(\mu_e)\bullet\left[(\mathrm{id}-P) - D\mathcal{F}(\mu_{2e},\lambda_c)\bullet P\right]\bullet(\delta\mu^*,\delta\mu^*) = 0.$$

Thus we can consider cases in which the primary branch is nonlinearly stable for $\lambda<\lambda_c$, or $\lambda>\lambda_c$ or both, for example we can have

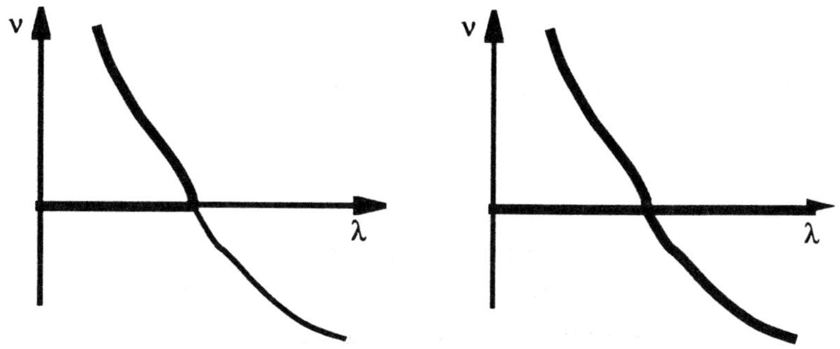

where the heavy curves done nonlinearly stable branches. However, the analysis does not apply at the point of bifurcation.

We have the following result when (4.3) is a differential equation:

THEOREM II: (Bifurcation Within a family)
Consider a Hamiltonian system which leads to a decomposable Arnold function. Suppose we have a branch of equilibria $\mu_e^a(x;\lambda)$ *depending on some parameter(s). Suppose further that the primary branch of equilibria is (nonlinearly) stable in the set* Λ_s *with* λ_c *in the closure of* Λ_s*, and that the equilibria in the neighborhood of* λ_c *are strictly monotonic in x. If the equation*

$$(\mathrm{id}-\mathrm{P})(\nu)-\left[\mathcal{F}\left(\mathrm{P}(\mu_e^a+\nu),\lambda\right)-\mathcal{F}\left(\mathrm{P}(\mu_e^a),\lambda\right)\right]=0,$$

together with homogeneous boundary conditions on ν*, undergoes a bifurcation from the trivial solution at* λ_c*, then the solution* $\mu_e^a(x;\lambda)+\nu(x;\lambda)$ *is nonlinearly stable for* $\lambda\in\Lambda_s$ *in the neighborhood of* λ_c*. Moreover, the bifurcating branch is nonlinearly stable for* $\lambda\in\Lambda_s$ *provided that*

$$\underset{D}{\mathrm{range}}\left[\mu_e^a(x;\lambda)\right]=\underset{D}{\mathrm{range}}\left[\mu_e^a(x;\lambda)+\nu(x;\lambda)\right]$$

remains satisfied.

We prove this result as follows. If the equilibrium $\mu_e^a(x;\lambda_c)$ is strictly monotonic in x, then sufficiently close to the bifurcation point $\mu_e^a(x;\lambda)+\nu(x;\lambda)$ is also monotonic in x. Because $\nu(x;\lambda)$ satisfies homogeneous boundary conditions, condition (4.2) holds. By hypothesis, $\mu_e^a(x;\lambda)$ and $\mu_e^a(x;\lambda)+\nu(x;\lambda)$ belong to the same family$(\mathcal{F},\mathrm{P})$. Thus, by Theorem I, nonlinear stability of μ_e^a

implies nonlinear stability of $\mu_e^b = \mu_e^a + v$ so long as equation (4.2) holds. In practice, this may persist for a large range of the parameter.

We remark on two importants strengths of this result. First, it does not depend on the type of bifurcation which equation (4.3) undergoes. The bifurcation may be highly degenerate, yielding multiple solution branches and even continua of solutions, which is an important and often *generic* possibility in systems with symmetry. In other approaches to this problem, inference of the stability results from a primary branch to a bifurcating branch is complicated by any degeneracy of the bifurcation. The second remark is that Theorem II may be applied to bifurcation to tertiary branches of equilibria, and so forth. This means that the stability of perhaps highly complicated branches of equilibria may be known *a priori* by a stability analysis of the underlying primary branch and by checking condition (4.2), a procedure which is generally much easier to handle than a new stability analysis.

Care must be exercised in the application of Theorem II to Hamiltonian systems where a wavenumber cut off is required to obtain stability results. Essentially, the problem is that one can apply "irrational" wavenumber cut offs in such a way as to render the second variation *definite* at a point of bifurcation, which does not make sense. However, it is possible to devise a cut off which respects the bifurcation, and thus apply Theorem II successfully. See the discussion in the example in section 5.

Note that one must carefully distinguish between a bifurcation of the equation $X_H(\mu_e)=0$ and a bifurcation of equation (4.3). The latter implies that the former occurs, but not vice versa; see Szeri (1989) for more details. In particular, the relationship between the eigenvalues of the linearized operator $DX_H(\mu_e)(x,\lambda)$ and the eigenvalues of $[(id-P)-D\mathcal{F}(P\mu_e,\lambda)\bullet P](x,\lambda)$, the linearization of equation (4.3), is not at all clear. In general, this relationship depends in a complicated way on the structure function $\mathcal{F}$ and the projection P.

Finally, we comment on an interesting possibility for specific examples of Hamiltonian systems to which one applies the methods we have outlined here. It may be possible to prove a result for specific systems which states that linear *instability* of (a part of) the primary branch of equilibria in a bifurcation problem implies linear instability of equilibria on the bifurcating branch at the same value of the parameter. This could be attempted using the structure function equation which links the two branches of equilibria and any of a number of classical techniques of spectral analysis of operators (see, for example, Drazin & Reid (1981)). Mathematically, the statement of this problem is to prove that if
there exists a $\sigma \in \text{spectrum}\{DX_H(\mu_e^a(x;\lambda),\lambda)\}$ with $\text{Re}(\sigma)>0$, then there exists a $\sigma' \in \text{spectrum}\{DX_H(\mu_e^b(x;\lambda),\lambda)\}$ with $\text{Re}(\sigma')>0$, where $\mu_e^a, \mu_e^b \in \text{family}(\mathcal{F},P)$. We have not yet found a general formulation of this relationship.

5. STABILITY AND BIFURCATION OF SWIRLING FLOWS. The class of axisymmetric swirling flows of an inviscid fluid is a Hamiltonian system with a noncanonical formulation given by S&H. In this section, we review the stability argument of sections three and four of that paper, recast in the present notation. There is, however, one difference between the present analysis and that of S&H. In order that we may obtain a second variation which "respects" bifurcation of the family equation (2.2), we perform the high wavenumber cut off differently, as we now explain.

In S&H, the high wavenumber cut off is applied to the operator in the first term of equation (2.12). In the example of swirling flows, $D^2H(\mu_e)$ turns out to contain an operator with minimum eigenvalue limiting on zero (from positive values), corresponding to an eigenfunction with arbitrarily large wavenumber. After cutting off the wavenumber of allowable disturbances, S&H replace the operator in $D^2H(\mu_e)$ by its minimum eigenvalue on the space of disturbances which satisfy the high wavenumber cut off; thus they replace $D^2H(\mu_e)$ by $D^2H(\mu_e)^{min}$. See

S&H for a discussion and physical justification of this step. Thus the second variation, after the cut off, is

$$\delta^2 H_{\hat{C}}(\mu_e) = D^2 H(\mu_e)^{min} \bullet (\delta\mu,\delta\mu)$$
$$- D^2 H(\mu_e) \bullet (\delta\rho + D\mathcal{F}(\rho_e,\lambda) \bullet \delta\rho,\delta\mu).$$

At the critical point where a bifurcation of the family equation (2.2) occurs, the second variation in the direction of bifurcation is found using equation (4.4), giving

$$\delta^2 H_{\hat{C}}(\mu_e) = \left[D^2 H(\mu_e)^{min} - D^2 H(\mu_e)\right] \bullet (\delta\mu, \delta\mu).$$

In the space of disturbances which satisfy the cut off, this expression is definite, and so does not make sense. Essentially, "bad" modes are removed from $D^2H(\mu_e)$ in one of its appearances and not in another. This casts serious doubt on the usefulness of the cut offs used in S&H and elsewhere. To rectify this situation, in the present analysis we replace $D^2H(\mu_e)$ by $D^2H(\mu_e)^{min}$ in the first *and second* terms of the second variation. Thus we take

$$\delta^2 H_{\hat{C}}(\mu_e) = D^2 H(\mu_e)^{min} \bullet (\delta\mu,\delta\mu)$$
$$- D^2 H(\mu_e)^{min} \bullet (\delta\rho + D\mathcal{F}(\rho_e,\lambda) \bullet \delta\rho,\delta\mu).$$

This change has the effect of yielding slightly more conservative stability criteria for swirling flows, but we do obtain a second variation which is indefinite at the point of bifurcation by this procedure.

Moving on to the formulation of the problem, we consider an inviscid fluid flowing through a cylindrical domain D of radius R and length L. Due to the assumption of axisymmetric motion, two dependent variables suffice to describe the motion; these are

$$\kappa = \text{swirl function} = r e_\theta \bullet u$$
$$\chi = \text{vortex density} = \frac{1}{r} e_\theta \bullet \text{curl} u. \tag{5.1a,b}$$

Here u is the velocity in cylindrical polar coordinates (r,θ,z). Related to the vortex density is an auxiliary variable, the Stokes stream function:

$$\chi = \mathcal{L}\psi = \left(-\frac{1}{2y}\frac{\partial^2}{\partial z^2} - \frac{\partial^2}{\partial y^2}\right)\psi, \tag{5.2}$$

where $y=r^2/2$ is a convenient radial coordinate. Corresponding to the abstract setup of section 2, we can choose either $\mu=(\chi,\kappa)$ or $\mu=(\psi,\kappa)$. It turns out that the former makes for more straightforward calculations.

The Hamiltonian function for this system is just the kinetic energy:

(5.3) $$H(\chi,\kappa)=\frac{1}{2}\int_D\left(\psi\chi+\frac{\kappa^2}{2y}\right)d^3x+\frac{Q}{2}\int_0^L w\Big|_R\,dz,$$

where Q is the flow rate through the cylindrical domain D, and w is the axial velocity. Using the Hamiltonian function and the noncanonical Poisson bracket derived in S&H, one can show that the evolution equations for χ and κ are

(5.4a,b) $$\frac{\partial\chi}{\partial t}+\{\psi,\chi\}=\left\{\kappa,\frac{\kappa}{2y}\right\}$$ $$\frac{\partial\kappa}{\partial t}+\{\psi,\kappa\}=0,$$

where

$$\{a,b\}=\frac{\partial a}{\partial y}\frac{\partial b}{\partial z}-\frac{\partial a}{\partial z}\frac{\partial b}{\partial y}$$

is the Jacobian. Equations (5.4) correspond to equation (2.1) in the abstract setup.

5.1 THE STRUCTURE FUNCTION. When evaluated at equilibrium, equations (5.4) lead to functional dependencies among the equilibrium fields $\kappa_e(y,z)$ and $\chi_e(y,z)$, or, equivalently, $\kappa_e(y,z)$ and $\psi_e(y,z)$. Evaluating (5.4) at an equilibrium and rearranging, yields

(5.5a,b) $$\{\psi_e,\kappa_e\}=0$$ $$\left\{\psi_e,\chi_e-\frac{1}{4y}\frac{d(\kappa_e)^2}{d\psi_e}\right\}=0.$$

Equations (5.5) are necessary and sufficient for the existence of two functional relationships:

(5.6a,b) $$\kappa_e=F(\psi_e)\quad\text{or}\quad\psi_e=K(\kappa_e)\quad\text{with}\quad K=F^{-1}$$ $$\chi_e-\frac{1}{4y}\frac{d(\kappa_e)^2}{d\psi_e}=V(\psi_e).$$

Thus, for this Hamiltonian system, the dependent variable is $\mu=(\chi,\kappa)$, and the projection P is defined by

$$P\mu_e=\begin{bmatrix}0 & 0\\ 0 & 1\end{bmatrix}\begin{pmatrix}\chi_e\\ \kappa_e\end{pmatrix}=\begin{pmatrix}0\\ \kappa_e\end{pmatrix} \tag{5.7a}$$

and

$$(\mathrm{id}-P)\mu_e=\begin{bmatrix}1 & 0\\ 0 & 0\end{bmatrix}\begin{pmatrix}\chi_e\\ \kappa_e\end{pmatrix}=\begin{pmatrix}\chi_e\\ 0\end{pmatrix}. \tag{5.7b}$$

The equation corresponding to the definition of the structure function (2.2) is therefore

$$\begin{pmatrix}\chi_e\\ 0\end{pmatrix}=\begin{pmatrix}\mathcal{F}_1(\kappa_e,\lambda)\\ \mathcal{F}_2(\kappa_e,\lambda)\end{pmatrix}. \tag{5.8}$$

Comparison with (5.6) shows

$$\mathcal{F}_1(\kappa_e,\lambda)=\mathcal{L}K(\kappa_e)$$
$$\mathcal{F}_2(\kappa_e,\lambda)=-\kappa_e+2\gamma K'(\kappa_e)\left[\mathcal{L}K(\kappa_e)-V(K(\kappa_e))\right]; \tag{5.9a,b}$$

note that at equilibrium, $\mathcal{F}_2=0$. Away from equilibrium, we have

$$\mathcal{F}(P\mu,\lambda)=\begin{pmatrix}\mathcal{L}K(\kappa)\\ -\kappa+2\gamma K'(\kappa)\left[\mathcal{L}K(\kappa)-V(K(\kappa))\right]\end{pmatrix}. \tag{5.10}$$

Note that if (5.8) is satisfied, then equations (5.6) are satisfied, which is a sufficient condition for (χ_e,κ_e) to be an equilibrium flow. Selecting a particular family means that we fix the functions V and K of (5.9a,b), exclusive of any parameter(s) we wish to keep free.

5.2 STABILITY ANALYSIS. The Arnold function is

$$H_c(\chi,\kappa)=\frac{1}{2}\int_D\left(\psi\chi+\frac{\kappa^2}{2\gamma}\right)d^3x+\int_D\left(j(\kappa)+\chi f(\kappa)\right)d^3x \tag{5.11}$$

where f and j are as yet undetermined real valued functions, and we have taken c of equation (3.1) of S&H to be $c=-Q/2\pi$ from the outset. We set the first variation evaluated at an equilibrium to be zero, and hence determine the functions j and f as in equation (2.8). This yields

$$f(\kappa)=-K(\kappa),$$
$$j'(\kappa)=-V(K(\kappa))\,K'(\kappa) \tag{5.12a,b}$$

and so

$$D\hat{C}(\kappa) \bullet (\delta\chi,\delta\kappa) = \int_D (-K(\kappa)\delta\chi - (\mathcal{L}K(\kappa) - V(K(\kappa)))\, K'(\kappa)\,\delta\kappa) d^3x. \tag{5.13}$$

The second variation evaluated at equilibrium is then

$$\delta^2 H_{\hat{C}}(\mu_e) = \int_D \delta\chi \mathcal{L}^{-1}\delta\chi - \delta\chi\,\delta\kappa K'(\kappa_e) - \delta\kappa K'(\kappa_e)\mathcal{L}(K'(\kappa_e)\,\delta\kappa)$$
$$+\left(\frac{1}{2y} + K''(\kappa_e)(V(\psi_e) - \chi_e) + V'(\psi_e)\,K'^2(\kappa_e)\right)(\delta\kappa)^2 d^3x.$$

Or, corresponding to the form of (2.12), we have

$$\delta^2 H_{\hat{C}}(\mu_e) = \int_D dx\,(\delta\chi,\delta\kappa)\begin{pmatrix}\mathcal{L}^{-1} & 0\\ 0 & \frac{1}{2y}\end{pmatrix}\begin{pmatrix}\delta\chi\\ \delta\kappa\end{pmatrix} - (\delta\chi, \delta\kappa)\begin{pmatrix}\mathcal{L}^{-1} & 0\\ 0 & \frac{1}{2y}\end{pmatrix} *$$

$$\begin{pmatrix}\mathcal{L}K'(\kappa_e)\,\delta\kappa\\ 2y\left(K''(\kappa_e)(\chi_e - V(\psi_e)) - V'(\psi_e)K'^2(\kappa_e) + K'(\kappa_e)\mathcal{L}K'(\kappa_e)\right)\delta\kappa\end{pmatrix}. \tag{5.14}$$

As discussed in the beginning of this section, we must cut off the wavenumber of disturbances in χ to avoid indefiniteness of the Arnold function to high frequency perturbations. Thus we replace the term $\mathcal{L}^{-1}$ in both matrices of (5.14) by $1/\lambda_N$, where

$$\lambda_N \approx \frac{4\pi^2 N^2}{R^4}$$

is an estimate of a high order eigenvalue of the operator $\mathcal{L}$. See S&H for more details and physical justification of this step.

The criterion of formal stability is the criterion for positive definiteness of (5.14) (after applying the cut off):

$$\frac{1}{2y} + K''(\kappa_e)(V(\psi_e) - \chi_e) + V'(\psi_e)\,K'^2(\kappa_e) > \frac{5}{4}\lambda_N K'^2(\kappa_e).$$

This is slightly more conservative than the criterion of S&H in which the factor 5/4 is replaced by 1. However, as discussed in the beginning of this section, the present implementation of the high wavenumber cut off yields a second variation which is indefinite at a point of bifurcation, as required. This can be seen by evaluating equation (5.14) in the direction of bifurcation.

Finally, we seek conditions for nonlinear stability. Corresponding to equation (2.14) we have

$$Q_1(\Delta\mu) = \int_D (\Delta\chi, \Delta\kappa)\begin{pmatrix} \mathcal{L}^{-1} & 0 \\ 0 & \frac{1}{2y} \end{pmatrix}\begin{pmatrix} \Delta\chi \\ \Delta\kappa \end{pmatrix} dx. \tag{5.15a}$$

Equation (2.16) is satisfied by

$$Q_2(\Delta\mu) = \int_D \left(\beta^-(\Delta\kappa)^2 + 2\alpha^+\Delta\chi\,\Delta\kappa\right) dx, \tag{5.15b}$$

where

$$\beta^- \le K''(\kappa_e)(V(\psi_e) - \chi_e) + V'(\psi_e)K'^2(\kappa_e) - \lambda_N K'^2(\kappa_e)$$

and

$$\frac{1}{2}\left|K'(\kappa_e)\right| \le \alpha^+ < \infty \ .$$

To satisfy (2.16), we require

$$\int_D \left(\Delta\chi\mathcal{L}^{-1}\Delta\chi + \frac{1}{2y}(\Delta\kappa)^2 + \beta^-(\Delta\kappa)^2 + 2\alpha^+\Delta\chi\Delta\kappa\right) dx > 0. \tag{5.16a}$$

Completing the square, we have

$$\int_D \left(\left(\sqrt{\mathcal{L}^{-1}}\Delta\chi + \frac{\alpha^+}{\sqrt{\mathcal{L}^{-1}}}\Delta\kappa\right)^2 + \left(\frac{1}{2y} + \beta^- - \frac{(\alpha^+)^2}{\mathcal{L}^{-1}}\right)(\Delta\kappa)^2\right) dx > 0.$$

This is true if

$$\frac{1}{R^2} + \beta^- > \lambda_N(\alpha^+)^2. \tag{5.16b}$$

The continuity of the norm is easily established. Also, one can see that the Arnold function for this example is decomposable.

5.3 BIFURCATION OF SWIRLING FLOWS. Because we have identified the Arnold function for the case of axisymmetric swirling flows as decomposable, Theorem II applies to this Hamiltonian system. Thus bifurcating branches of the structure function equation will be stable near the bifurcation point in the range of the parameter where the primary branch is stable. The stability of these bifurcating branches holds for at least the range of the parameter in which solutions of the bifurcating branch are monotonic and the primary branch is stable.

We will set up the bifurcation problem for a family of equilibria which depends on a real parameter β via the multiplicative relationship

(5.17) $\quad F(\psi) = \beta f(\psi),$

for some fixed f, with corresponding changes in the structure function, etc. It turns out to be more convenient to work in terms of F rather than K. The bifurcation problem corresponding to equation (4.3) is

$$\begin{pmatrix}\chi\\0\end{pmatrix} - \begin{pmatrix}\mathcal{F}_1(\kappa_e+\kappa,\beta) - \mathcal{F}_1(\kappa_e,\beta)\\ \mathcal{F}_2(\kappa_e+\kappa,\beta) - \mathcal{F}_2(\kappa_e,\beta)\end{pmatrix} = 0. \tag{5.18}$$

Here $\mu_e^a = (\chi_e^a, \kappa_e^a)$ is the primary branch, and $\mu_e^b = (\chi_e^a + \chi, \kappa_e^a + \kappa)$ is the bifurcating branch; we also have the Stokes stream function on the primary branch, ψ_e^a, and on the bifurcating branch, $\psi_e^a + \psi$. Using the definition of the structure function, equation (5.10), equation (5.18) becomes

(5.19a) $\quad \chi_e^a + \chi = \mathcal{L}F^{-1}(\kappa_e^a + \kappa)$

and

$$\mathcal{L}\psi + \chi_e^a - V(\psi_e^a + \psi) - \frac{1}{2y}F(\psi_e^a + \psi)F'(\psi_e^a + \psi) = 0. \tag{5.19b}$$

Note that the bifurcation equations are just statements that the new branch of equilibria will share the same functions F (or K) and V as the primary branch of equilibria. The boundary conditions for equations (5.19) are that ψ vanishes on $r=0$ and $r=R$, and ψ is periodic of period L in z.

We obtain an infinitesimal version of the bifurcation problem by expanding

$$\psi = \varepsilon\tilde{\psi} + \mathcal{O}(\varepsilon^2)$$

and substituting this expansion into equation (5.19b). Differentiating by ε and setting $\varepsilon=0$ yields

$$\mathcal{L}\tilde{\psi} - \left(V'(\psi_e^a) + \frac{1}{2y}\left(F'^2(\psi_e^a) + F(\psi_e^a)F''(\psi_e^a)\right)\right)\tilde{\psi} = 0. \tag{5.20}$$

The infinitesimal problem may be simplified in the case when $\mu_e^a = (\chi_e^a, \kappa_e^a)$ is a branch of *columnar* equilibria. Columnar equilibria are characterized by translation invariance of the equilibrium in the axial direction; thus we have $\kappa_e^a = \kappa_e^a(y)$, $\psi_e^a = \psi_e^a(y)$ and $\chi_e^a = \chi_e^a(y)$. In addition, because we are interested in branches of strictly monotonic equilibria, we can invert

$\psi_e^a = \psi_e^a(y)$ to obtain $y = Y(\psi_e^a)$. For a columnar primary branch, (5.20) may be simplified to the form

$$\mathcal{L}\tilde{\psi} - \left(\frac{d\chi_e^a}{d\psi_e^a} + \frac{F(\psi_e^a)F'(\psi_e^a)}{2y^2 \frac{d\psi_e^a}{dy}} \right) \tilde{\psi} = 0. \tag{5.21}$$

5.4 EXAMPLE. Finally, we come to an example of a bifurcation from one axisymmetric swirling flow to another. The primary branch consists of a flat axial velocity profile, together with a Burgers vortex swirl velocity profile: specifically, we take

$$\psi_e^a(y) = y, \quad \text{and} \quad \kappa_e^a(y) = \beta(1 - e^{-\alpha y}). \tag{5.22a,b}$$

We also choose $\alpha > 0$ fixed and let $\beta > 0$ be the variable bifurcation parameter. Note that the primary branch of columnar flows are of the form (5.17). It is easy to compute

$$F(\tau) = \beta f(\tau) = \beta(1 - e^{-\alpha\tau}) \tag{5.23a}$$

and

$$V(\xi) = -\beta^2 \left(\frac{f(\xi)f'(\xi)}{2\xi} \right). \tag{5.23b}$$

The infinitesimal version of the bifurcation problem is

$$\mathcal{L}\tilde{\psi} - \beta^2 \left(\frac{\alpha e^{-\alpha y}(1 - e^{-\alpha y})}{2y^2} \right) \tilde{\psi} = 0. \tag{5.24}$$

The boundary conditions are $\psi(r=0) = 0 = \psi(r=R)$ and ψ is L-periodic in z.

Equation (5.24) is solved by separation of variables, by which one finds that

$$\tilde{\psi} = Y_m(y) \cos\left(\frac{2n\pi}{L} z \right). \tag{5.25}$$

Here, $n = 0, 1, 2\ldots$, and the y-component is given by

$$Y''_m(y) + \left(\frac{\alpha\beta_{nm}^2 e^{-\alpha y}(1 - e^{-\alpha y})}{y} - \frac{4n^2\pi^2}{L^2} \right) \frac{Y_m(y)}{2y} = 0 \tag{5.26}$$

with $Y(0) = Y(1/2) = 0$, and we have set $R = 1$. This two-point boundary value problem is a linear eigenvalue problem for the pair $(\beta_{nm}, Y_m(y))$, where β_{nm} is the critical value of the swirl parameter corresponding to (5.25), which has n internal zeros in the z-direction, and m internal zeros in the y-direction. We solved this problem using AUTO, a software package developed for continuation and bifurcation problems in ordinary differential equations; see Doedel (1986) for descriptions of the algorithms and implementation. A list of the first few critical swirl values appears in table 1.

TABLE 1

n	m	β_{nm}
0	0	0.4323
	1	0.8334
	2	1.361
1	0	0.5834
	1	0.9772
	2	1.570
2	0	0.7954
	1	1.203
	2	1.904

Next, we chose to continue (numerically) the solution which bifurcates at the critical swirl value β_{00}=0.4323. This is a bifurcation to another columnar, or z-independent flow, which shares the same functions F and V as the primary branch of equilibria. The full nonlinear bifurcation equation for $\psi(y;\beta)$ is

$$\frac{d^2\psi}{dy^2}+\frac{\alpha\beta^2 e^{-\alpha(y+\psi)}\left(1-e^{-\alpha(y+\psi)}\right)}{2y}\left(\frac{\psi}{y+\psi}\right)=0 \tag{5.27}$$

with the boundary conditions $\psi(0)=\psi(1/2)=0$. Again using AUTO, we trace out the bifurcation diagram of figure 1.

The bifurcating branch is characterized by an axial jet ($\beta>\beta_{00}$), or an axial wake ($\beta<\beta_{00}$) structure. The lower limit of monotonicity of $\psi_e^b(y)=y+\psi$, is β=0.3390. Although we continued the bifurcating branch to β=15, we never reached the upper limit of monotonicity! We show pictures of various equilibria on the bifurcating branch below.

A stability analysis of the primary branch yields the criterion

$$\beta^2 > \frac{5\pi^2}{2} N^2 \frac{e^\alpha}{\alpha\left(e^{\alpha/2} - 1\right)}. \tag{5.28}$$

Equation (5.28) corresponds to equation (5.16b). This criterion yields hopelessly small wavenumber cut offs for low values of the swirl parameter β. However, we have chosen this example for demonstrative, rather than for practical purposes. With this in mind, we can choose a (very) small wavenumber cut off such that $\beta > 0.3$, say, is a stability criterion for the primary branch. Then by Theorem II, the secondary branch is also stable, with identical wavenumber cut off, in the range of the parameter where the equilibria on the secondary branch are monotonic, that is $\beta > 0.3390$ to at least $\beta = 15$.

Finally, we show two equilibria on the bifurcating branch which fall into this range. The first is computed at $\beta = 0.34$, near the lower limit of monotonicity. The perturbation stream function is shown in figure 2a. In figure 2b is a plot of the axial velocity, which includes the axial velocity of the support flow. One can see the strong wake-like structure near the axis of the coordinate system. Note that these plots are drawn with respect to the true radial coordinate r and not y. In figures 3a and 3b, we repeat these pictures for an equilibrium computed at $\beta = 2$, that is on the jet-like side of the bifurcation point. Here the jet is already very strong and concentrated near the axis, with a center line velocity of nearly 54! We are certainly "far" from the bifurcation point; nevertheless, Theorem II applies to this case as it does to the flow in figures 2a and 2b.

6. CONCLUSIONS. Through exploitation of the natural structure of functional dependence among the equilibrium fields of a Hamiltonian system, we have been able to link the ideas of nonlinear stability and bifurcation. We began by postulating the

functional dependence at equilibrium in the form of a projection operator and the structure function. Next, we used the energy-Casimir method of Arnold to obtain sufficient conditions for nonlinear stability of equilibria of the system.

We observed that, for Hamiltonian systems which lead to decomposable Arnold functions, stability results may be extended from one equilibrium within a family to another, provided the ranges of the dependent variables match. Thus stability of bifurcating branches may be inferred from that of the primary branch, perhaps for large ranges of the parameter. Moreover, this result is applicable despite any degeneracies in the bifurcation itself.

Finally, we treated a particular noncanonical Hamiltonian system: the axisymmetric swirling flows of an inviscid fluid. Using the energy-Casimir method, we obtained stability criteria for this system. Next, we analyzed a bifurcation from a simple branch of equilibria to a branch of sufficient complexity that numerical computation of the equilibria was necessary. We found the limits of monotonicity of the stream function on the bifurcating branch, and therefore the range of applicability of Theorem II.

Although we devoted more attention to Theorem II, Theorem I has potentially wider application. For instance in the example of section 5.4, we were able to apply Theorem II for β up to (and beyond) $\beta=15$. However, the cut off is irrationally small at the point of bifurcation. We could instead apply Theorem I at $\beta=15$, which would imply stability of the secondary branch to disturbances having a much more sensible cut off. From this point of view, the bifurcation results are only necessary to establish that the two branches belong to the same family.

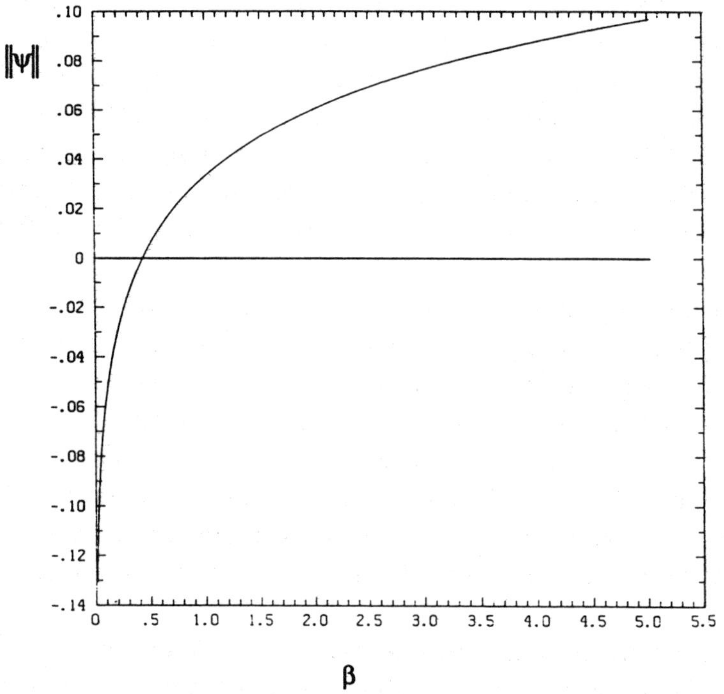

Figure 1. Partial bifurcation diagram of equation (5.27). Plotted are twice the integral of ψ versus the parameter β. The region of monotonicity of equilibria on the bifurcating branch is $\beta>0.3390$.

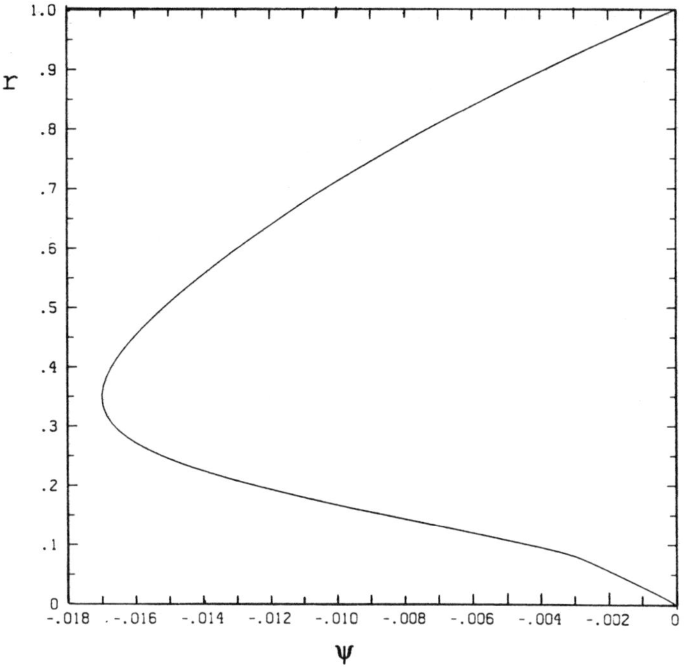

Figure 2a. Perturbation stream function as a function of r for an example of an equilibrium on the bifurcating branch of figure 1 at $\beta=0.34$.

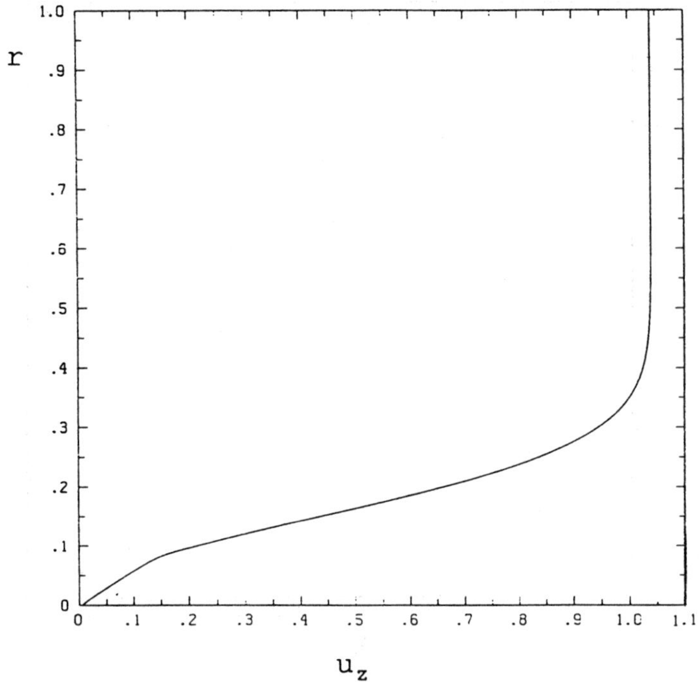

Figure 2b. Total Axial velocity as a function of r for the equilibrium of figure 2a.

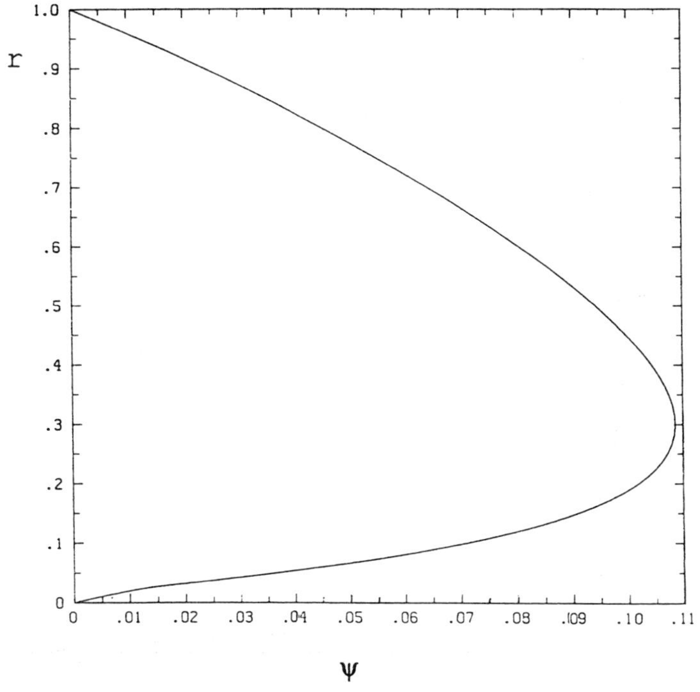

Figure 3a. Perturbation stream function as a function of r for an example of an equilibrium on the bifurcating branch of figure 1 at $\beta=2$.

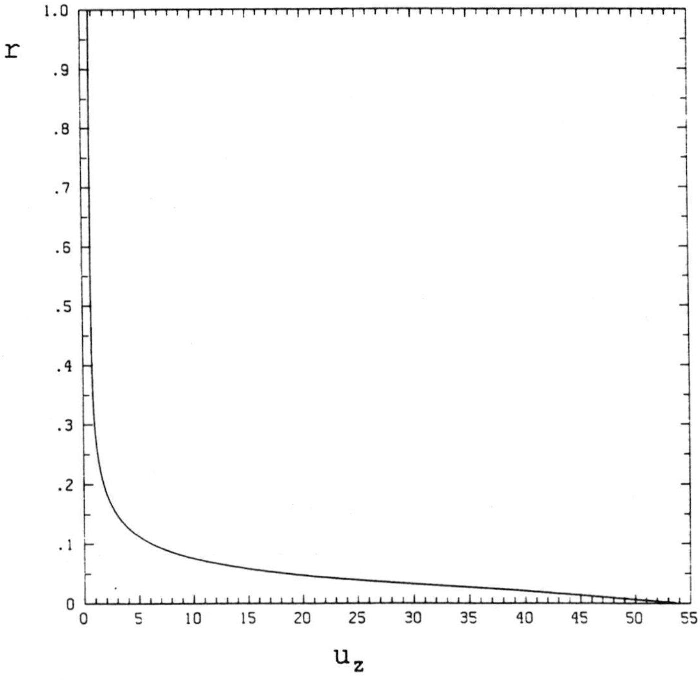

Figure 3b. Total Axial velocity as a function of r for the equilibrium of figure 3a.

BIBLIOGRAPHY

1. Abarbanel, H. D. I., Holm, D. D., Marsden, J. E. & Ratiu, T. S., "Nonlinear stability of stratified fluid equilibria", Phil. Trans. R. Soc. Lond., A 318 (1986), 349 - 409.

2. Arnold, V. I., "Conditions for the nonlinear stability of the stationary plane curvilinear flows of an ideal fluid", Dokl. Mat. Nauk., 162 (1965), 773 - 777.

3. Arnold, V. I., "On an a priori estimate in the theory of hydrodynamic stability", [English translation] Am. math. Soc. Transl., 19 (1969), 267 - 269.

4. Batchelor, G. K., "On steady laminar flow with closed streamlines at large Reynolds number", J. Fluid Mech., 1 (1956), 177 - 190.

5. Benjamin, T. B., "Theory of the vortex breakdown phenomenon", J. Fluid Mech., 14 (1962), 503-529.

6. Benjamin, T. B., "Impulse, flow force and variational principles", IMA J. Appl. Maths., 32 (1984), 3-68.

7. Doedel, E., AUTO: Software for continuation and bifurcation problems in ordinary differential equations. Report, Applied Mathematics, California Institute of Technology, Pasadena, 1986.

8. Drazin, P. G. and Reid, W. H., Hydrodynamic Stability, Cambridge University Press, New York, 1981.

9. Golubitsky, M. and Schaeffer, D. G., Singularities and groups in bifurcation theory. Volume I., Springer - Verlag, New York, 1985.

10. Golubitsky, M. and Schaeffer, D. G., Singularities and groups in bifurcation theory. Volume II., Springer - Verlag, New York, 1988.

11. Holm, D. D., Marsden, J. E., Ratiu, T. S. & Weinstein, A., "Nonlinear stability of fluid and plasma equilibria", Physics Rep., 123 (1985), 1 - 116.

12. Leibovich, S., "Weakly nonlinear waves in rotating fluids", J. Fluid Mech., 42 (1970), 803 - 822.

13. Leibovich, S., "Fully nonlinear structures, wavetrains and solitary waves in vortex filaments", in Nonlinear wave interactions in fluids, ed. by Miksad, R. W., Akylas, T. R. & Herbert, T., American Society of Mechanical Engineers, New York, 1987.

14. Lewis, D., "Nonlinear stability of a rotating liquid drop", Arch. ration. Mech. Analysis., (1989), in press.

15. Squire, H. B., "Rotating Fluids", in Surveys in Mechanics, ed. by G. K. Batchelor and R. M. Davies., Cambridge University Press, London, 1956.

16. Szeri, A. J. & Holmes, P. J., "Nonlinear stability of axisymmetric swirling flow", Phil. Trans. R. Soc. Lond., A 326 (1988), 327-354.

17. Szeri, A. J., "Nonlinear stability and bifurcation of fluid and plasma equilibria", (1989), in preparation.

ACKNOWLEDGEMENTS

The authors would like to thank Debbie Lewis and Juan Simo for a helpful discussion. We are grateful to Jerry Marsden for his continuing interest, encouragement and advice.

Applied Mechanics, 104-44
California Institute of Technology
Pasadena, California 91125.

Department of Mathematics and
Department of Theoretical and Applied Mechanics
Cornell University, Ithaca, New York 14853.

Contemporary Mathematics
Volume **108**, 1990

NONLINEAR RESONANCE IN INHOMOGENEOUS SYSTEMS OF CONSERVATION LAWS[1,2]

Eli Isaacson[3] and Blake Temple[4]

ABSTRACT: We solve the Riemann problem for a general inhomogeneous system of conservation laws in a region where one of the nonlinear waves in the problem takes on a zero speed. We state generic conditions on the fluxes that guarantee the solvability of the Riemann problem, and these conditions determine a unique underlying structure to the solutions. The inhomogeneity is modeled by a linearly degenerate field. Our analysis thus provides a general framework for studying (what we are calling) resonance between a linear and a nonlinear family of waves in a system of conservation laws. Special cases of this phenomenon are observed in model problems for gas dynamical flow in a variable area duct and in Buckley-Leverett type systems that model multiphase flow in a porous medium.

§1 INTRODUCTION: When two different families of waves take on the same wave speed in a nonlinear problem, we say that nonlinear resonance occurs [11,12]. When wave speeds from different families are not distinct, the number of times a pair of waves will interact cannot be bounded apriori. Consequently, since waves are reflected

---

[1] *Mathematics Subject Classification* (1985 Revision): 35L65

[2] The final version of this paper will be submitted for publication elsewhere.

[3] Supported in part by the National Science Foundation and by the Institute of Theoretical Dynamics, UC-Davis, 95616

[4] Supported in part by the National Science Foundation, Grant No. DMS-86-13450 and by the Institute of Theoretical Dynamics, UC-Davis,95616

in other families every time a pair of waves interact, a proliferation of reflected waves can occur by the interaction of a single pair of waves. Here we introduce a general framework in which resonant interaction between a linear and a nonlinear family of waves takes place. Such resonance arises in an inhomogeneous system of conservation laws when a nonlinear family of waves takes on a zero speed. By an inhomogeneous system of conservation laws we mean a system of the general form

$$u_t + f(a,u)_x = 0, \tag{1.1}$$

where we assume that $a=a(x)$ is a variable function of x alone, so that a represents an inhomogeneity in the problem. We express this by the additional conservation law

$$a_t = 0. \tag{1.2}$$

(Systems of this form were previously identified by the authors when they outlined a program for classifying the solutions of nonstrictly hyperbolic systems. [cf. 5,8]) Our general problem thus becomes

$$U_t + F(U)_x = 0, \tag{1.3}$$

where $U=(a,u)$, $F(U)=(0,f(a,u))$, and $u=(u_1,u_2,\ldots,u_n)\varepsilon R^n$, $x\varepsilon R$, $t\geq 0$. System (1.3) is a system of $n+1$ equations in the $n+1$ unknowns a, $u_1,\ldots,u_n$. We assume that for each fixed value of a, system (1.1) is a strictly hyperbolic system of n equations, and that each of the characteristic fields associated with the u variables is either genuinely nonlinear or linearly degenerate [10]. Equation (1.2) produces a linearly degenerate field in system (1.3) with eigenvalue $\lambda_0=0$ and corresponding eigenvector R_0 (i.e., $\nabla\lambda_0\cdot R_0=0$). The remaining eigenvalues,

$$\lambda_1 < \lambda_2 < \ldots < \lambda_n,$$

of system (1.3) correspond to the eigenvalues of system (1.1) and have corresponding eigenvectors $R_1,\ldots,R_n$ which lie in the hyperplane a=const. We wish to study this system in the neighborhood of a state $U_*=(a_*,u_*)$ at which a nonlinear family of waves in system (1.3) takes on a zero wave speed. Thus we assume that

$$\lambda_k(U_*)=\lambda_0=0, \tag{1.4}$$

and that

$$\nabla\lambda_k \cdot R_k \neq 0. \tag{1.5}$$

In §3 we state a theorem which gives generic conditions on the flux function F that guarantee the unique solvability of the Riemann problem in a neighborhood of a state U_* at which (1.4) and (1.5) hold. The Riemann problem for (1.3) is the initial value problem for piecewise constant initial data

$$U(x,0) = \begin{cases} U_L & \text{for } x < 0, \\ U_R & \text{for } x > 0. \end{cases}$$

The Riemann problem is fundamental to the study of (1.3) because it identifies the elementary waves that propogate, and these are typically shock waves, rarefaction waves and contact discontinuities. The conditions of our theorem determine a unique underlying structure to the solution of the Riemann problem in a neighborhood of U_*.

Special cases of (1.4) and (1.5) are observed in model problems for gas dynamical flow in a variable area duct and in Buckley-Leverett type systems that model multiphase flow in a porous medium. In the latter case the model equations are not given in the form of an inhomogeneous system of conservation laws, but we show that there exists a Langrangian type transformation that maps the given equations to equivalent systems (in the weak sense) that are of this form. (The transformation was shown to the authors by D. Marchesin and Jorge Patino.) Examples of such systems have been studied by Keyfitz and Kranzer [9] and by the authors [4,16], and under the Lagrangian transformation these turn out to be equivalent to an inhomogeneous scalar conservation law in our theory. Many but not all of the features in the scalar case carry over to the case of an inhomogeneous system of conservation laws. For example, there are in general $n+2$ waves in a solution even though there are only $n+1$ equations; Riemann problem solutions depend continuously on the data in xt-space but not in state space; but unlike the scalar case, the wave curves in the case of systems are only Lipschitz continuous curves near the point of resonance. This makes it difficult to apply the implicit function theorem directly, and we show that the existence and uniqueness of solutions of the Riemann problem in a neighborhood of a point of resonance is a consequence of the uniqueness of intersection points of Lipschitz continuous manifolds of complementary dimension. Our goal is to obtain an existence theory for the Cauchy problem using Glimm's method that applies in a neighborhood of a point of resonance in a general inhomogeneous system of conservation laws. Such a theorem in the scalar case can be obtained by methods introduced earlier by the second author [16], but a sharper bound on the total variation of solutions, as well as a quadratic potential interaction functional is required to generalize these methods to systems. Such a quadratic functional has not been found in any other case in which there is no apriori bound on the number of times a pair of waves will interact. In the scalar case the authors recently identified the asymptotic wave structure of solutions as t tends to infinity. Indeed, solutions in general decay to *inadmissible* solutions of the Riemann problem, and as a consequence of this, a lack of continuous dependence on the data in the L^p sense was observed [6]. The authors conjectured that some form of continuous dependence would be retrieved when viscosity effects were included. The analysis gives an explanation of how nonuniqueness of solutions of the Riemann problem is explained in terms of the time dynamics of general solutions, and similar phenomena occur in the case of an inhomogeneous system of conservation laws. The argument for decay in [6] went as follows: In [16] the second author constructed a positive nonincreasing functional defined on solutions at every time t ; and in [6] the authors showed that this functional was minimized on the asymptotic waves patterns among all possible wave patterns that a given solution could take on. The authors believe that a quadratic functional that succeeds for Glimm's method in an inhomogeneous system with resonance would help in completeing the proof of decay and would shed light on the rate of decay in the scalar case. Further directions include generalizing to the case where $a(x)$

is a vector. The authors recently analysed this problem in a case equivalent to the case when u is a scalar, but in the context of the multiphase flow problem [4,16].

§2 **APPLICATIONS:** In this section we describe two settings in which resonance in inhomogeneous systems of conservation laws arise.

Flow in a variable area duct: The equations for gas dynamical flow in a variable area duct with cross-sectional area $a(x)$ are [1]

$$\begin{aligned} &\rho_t + (\rho u)_x = -(a'/a)\rho u, \\ (1.6) \qquad &(\rho u)_t + (\rho u^2+p)_x = -(a'/a)\rho u^2, \\ &E_t + [(E+p)u]_x = -(a'/a)[(E+p)u]. \end{aligned}$$

We say that resonance occurs in transonic flow because then one of the nonlinear waves in the problem can be zero [cf 11]. Liu was the first to study the initial value problem for these equations from the point of view of Glimm's Random Choice Method, and he proved convergence of the method for solutions taking values in a neighborhood of a state $(\rho,\rho u,E)$ at which none of the wave speeds in the problem is zero (see [11] and references therein) . In [12] Liu also gave a fairly complete analysis of a scalar model for (1.6) in which resonance occurs. For systems, however, there is at present no general proof that Glimm's method converges in the transonic regime. To study this case, we rewrite these equations in the form

$$\begin{aligned} &(a\rho)_t + (a\rho u)_x = 0, \\ (1.7) \qquad &(a\rho u)_t + (a\rho u^2+ap)_x = -a'p, \\ &(aE)_t + [a(E+p)u]_x = 0, \end{aligned}$$

with the supplementary equation

$$(1.8) \qquad a_t=0.$$

The equations obtained when the zero order term on the right hand side is dropped yield a mathematical model for the resonant behavior that occurs in the transonic flow. The resulting system falls into our class. Note that the reduced system can be viewed also as the first system to solve in a numerical time splitting method for solving the original problem (1.7). In the special case that $p=c^2\rho$ (isothermal flow), the energy equation drops out and the zero order term can be incorporated into the fluxes to obtain the system

$$a_t = 0,$$

$$(1.9) \qquad (a\rho)_t + (a\rho u)_x = 0,$$

$$u_t + (u^2/2 + c^2 \log \rho)_x = 0.$$

Although this does not supply a physical conservation form for the original problem, it does provide a mathematical model containing a similar nonlinear resonance. For flow in a variable area duct, we believe that these models isolate an important component in the complicated behavior of transonic flow. Marchesin and Paes-Leme [13] studied this system in an analysis of the Riemann problem obtained by taking a to be piecewise constant, and our point of view here was influenced significantly by their analysis.

Buckley-Leverett type systems: We call the following equations the polymer equations because they arise as a model for the polymer flooding of an oil reservoir; i.e., a two phase, three component flow in a porous medium [3,16]:

$$s_t + f(s,c)_x = 0,$$

$$(1.10)$$

$$(cs)_t + (cf(s,c))_x = 0.$$

Here s and c correspond to a saturation and a concentration, resp., $0<s,c<1$, $f=f(s,c)$ is a constitutive relation, and the structure of solutions is determined by qualitative properties of f [4,16]. The eigenvalues of system (1.10) coincide when $f_s = f/s$. The Riemann problem for this system was studied by Isaacson in [4], and Keyfitz and Kranzer [9] studied the Riemann problem for the system

$$
\begin{aligned}
&u_t + [ug(u,v)]_x = 0, \\
(1.11)\qquad & \\
&v_t + [vg(u,v)]_x = 0,
\end{aligned}
$$

which is formally equivalent to the polymer system, and arose in their study of elasticity. The polymer interpretation of these equations suggests a natural Lagrangian transformation of the variables. In this model $g=f/s$ is the particle velocity of the water, and so the particle trajectories are given by solutions of the ordinary differential equation

$$x' = g(s(x,t),c(x,t)).$$

We can thus define a solution dependent mapping of the independent variables (x,t) to (ξ,t) so that ξ=const. defines the particles trajectories in the transformed, or Lagrangian coordinates ξ and t. One can verify that this is implemented through the mapping defined by specifying $x(\xi,t)$ through

$$\frac{\partial x(\xi,t)}{\partial t} = g(x(\xi,t),t) \ ,$$

$$x(\xi,0) = \int_0^{\xi} \frac{1}{s_0(x)}\, dx \ .$$

Rewriting system (1.10) in the ξt-coordinates yields the equivalent system

$$
\begin{aligned}
&c_t = 0, \\
(1.13)\qquad & \\
&(1/s)_t - g(s,c)_\xi = 0,
\end{aligned}
$$

which is a system of form (1.1) , (1.2) when we make the identifications $u=1/s$, $a=c$ and $h = -g$. Systems (1.10) and (1.13) are equivalent in the sense that they determine the same weak solutions under the 1-1 mapping given by Lagrangian change of variables. Furthermore, system (1.13) satisfies the assumptions of our theorem at the points where $\lambda_0=\lambda_1$. Thus writing $a=a(\xi)$, system (1.13) is an example of a scalar inhomogeneous equation in our framework.

§3 THE RIEMANN PROBLEM: We consider the system of equations

$$a_t = 0,$$

$$u_t + f(a,u)_x = 0,$$

where $u=(u_1, u_2, \ldots, u_n)$ and $a \in R$. We can restate this in the form (1.3) by taking $U=(a,u)$ and $F=(0,f)$. Here, $a=a(x)$ is an inhomogeneity in the equations, and $a_t=0$ gives rise to a linearly degenerate field with wave speed $\lambda_0=0$. We consider the Riemann problem for weak solutions in a neighborhood of a state $U_*=(a_*, u_*)$ at which

$$\lambda_1 < \ldots < \lambda_k = \lambda_0 < \ldots < \lambda_n.$$

This represents the simplest example of a coincidence of wave speeds.

THEOREM: *Assume that* f *satisfies the following conditions in a neighborhood of a state* $U_*=(a_*,u_*)$:

(i) *For each fixed value of* a, *the system*

$$\text{(u)} \qquad u_t + f(a,u)_x = 0$$

is strictly hyperbolic and genuinely nonlinear with eigenvalues

$$\lambda_1 < \lambda_2 < \ldots < \lambda_n.$$

(ii) $\lambda_k(a_*,u_*) = 0$.

(iii) *The* $n\times(n+1)$ *matrix* $\partial f/\partial U$ *has full rank at* U_*.

(iv) *The directional derivative of* a_0 *in the direction* R_0 *satisfies*

$$\nabla a_0 \cdot R_0 \big|_{U_*} \neq 0,$$

where R_0 *is the unit eigenvector for* $\lambda_0=0$ *and* a_0 *is the a-component of* R_0, *a function of* U.

Under assumptions (i)-(iv), *there exists a unique solution of the Riemann problem in a neighborhood of* U_*, *and in physical space this solution depends continuously on the left and right states. Moreover, for every* f *in this class, the solutions exhibit the same qualitative behavior.*

We indicate the proof here. For details the reader is referred to our forthcoming paper. First of all, the assumption of genuine nonlinearity $(\nabla\lambda_k \cdot r_k \neq 0)$ in the k'th field of system (u) guarantees that the equation $\lambda_k=0$ defines a smooth n-dimensional surface locally in R^{n+1}, passing through the state U_*. We call this the transition surface $\mathcal{T}$. By (i), $R_k=(a_k,r_k)$ points along the surface

$a=\text{const.}$, and the condition $\nabla\lambda_k \cdot r_k \neq 0$ also guarantees that the integral curves of R_k cut the transition surface transversally at $\mathcal{T}$. By (ii) and (iii), $\partial F/\partial U$ has Jordan normal form

$$
\text{(J)} \qquad \begin{bmatrix} \lambda_1 & & & & & & & \\ & \ddots & & & & & 0 & \\ & & \lambda_{k-1} & & & & & \\ & & & 0 & 1 & & & \\ & & & 0 & 0 & & & \\ & & & & & \lambda_{k+1} & & \\ & 0 & & & & & \ddots & \\ & & & & & & & \lambda_n \end{bmatrix}
$$

at U_* because $\partial F/\partial U$ has (full) rank n and $\lambda_0=\lambda_k=0$ at U_*. Moreover, $\partial F/\partial U$ has this normal form for every $U \in \mathcal{T}$ in a neighborhood of U_* since (iii) is an open condition. In particular, this implies that the eigenvectors R_0 and R_k agree on $\mathcal{T}$, and a_0, the 0'th component of R_0, vanishes on $\mathcal{T}$ in this neighborhood. Thus we can conclude that the integral curves for both R_0 and R_k cut the surface $\mathcal{T}$ transversally near the state U_*. Finally, the genericity condition (iv), that $\nabla a_0 \cdot R_0 |_{U_*} \neq 0$, implies that the integral curves of R_0 do not cross the surface $a=a_*$ at states U near U_*, and that the integral curve of R_0 passing through U_* must cross the surface $a=\text{const.}$ exactly twice at values of a near $a=a_*$, $a < a_*$. Indeed, a as a function of arclength along the integral curve of R_0 would have an inflection point at U_* if this integral curve stayed within or crossed the surface $a=a_*$ at U_*; and this would imply that $\nabla a_0 \cdot R_0 = 0$ on $\mathcal{T}$, violating (iv). We assume without loss of generality that the integral curve of R_0 lies below the surface $a=a_*$ near the state U_* (see Figure 1). Since all of the conditions (i) through (iv) are open conditions, the above conclusions about the integral curve of R_0 through U_* must also hold for all $U \varepsilon \mathcal{T}$ in a neighborhood of U_*.

The solution of the Riemann problem for arbitrary states U_L and U_R in a neighborhood of U_* is constructed as follows: we let

$T^i{}_t(U_L)$ denote the state t arclength units from U_L along the i-wave curve of U_L, $i=1,\dots,n$. (The i-wave curve of U_L consists of all right states that can be connected to U_L by an admissible i-wave, [10].) By (i), all states in the image of $T^i(U)$ lie at level a_L. For a given value of a_R, let $T^R(U_L)$ denote the set of all right states at level a_R that can be connected to U_L by a solution of the Riemann problem consisting of admissible k-waves and 0-waves alone; and let $T^R{}_t(U_L)$ denote the point t arclength units from the transition surface along $T^R(U_L)$. (Choose t to increase in the direction of λ_k.). We say that a 0-wave which connects U_L to U_R on the same integral curve of R_0 by a contact discontinuity of speed zero is *admissible* if the integral curve of R_0 does not cross the transition surface $\mathcal{T}$ between U_L and U_R. (Admissibility here is equivalent to conservation of the total variation of a in Glimm's method [cf 4,9,16]). The curves $T^R(U_L)$ are skethed in figures 2 and 3. Note that $T^R(U_L)$ is a continuous curve at level a_R, but is only Lipschitz continuous due to a possible jump in the derivative at the points labeled Q in figures 2 and 3. The continuity of the curves $T^R(U_L)$ at the special points Q follows from the triple jump condition formulated in [5]. The solution of the Riemann problem for arbitrary U_L and U_R is constructed by finding $t_1,\dots,t_n$ such that

$$U_R = T^n{}_{t_n} \cdots T^{k+1}{}_{t_{k+1}} \cdot T^R{}_{t_k} \cdot T^{k-1}{}_{t_{k-1}} \cdots T^1{}_{t_1}(U_L).$$

By definition the elementary waves corresponding to the $T^i{}_{t_i}(U_i)$ take U_L to U_R as i ranges from 1 to n, and this determines the unique solution of the Riemann problem near U_*. Since $T^R(U)$ is only Lipschitz continuous, the implicit function theorem is difficult to apply directly to obtain existence and uniqueness of $t_1,\dots,t_n$ for each pair U_L and U_R in a neighborhood of U_*. Existence and uniqueness is verified by demonstrating the uniqueness of intersection points for Lipschitz continuous manifolds of conplementary dimension. The procedure goes as follows. We first make the definition

DEFN: A function

$$\phi : \{t \varepsilon \mathbf{R}^k \ni |t| < \tau\} \to R^n$$

defines a Lipschitz continuous manifold with ε-approximate tangent vectors $W_1,...,W_n$ in R^n if

$$\left| \frac{\phi(\mathbf{t}+\alpha \mathbf{e}_i)-\phi(\mathbf{t})}{\alpha} - W_i \right| < \varepsilon$$

whenever $|\mathbf{t}|<\tau$ and $|\mathbf{t}+\alpha \mathbf{e}_i|<\tau$. It is then not hard to prove the following lemma:

LEMMA 1: Let $\phi = \phi(\mathbf{t_k})$ and $\varphi(\mathbf{t_{n-k}})$ define Liplchitz continuous manifolds M^k and N^{n-k} with ε-approximate tangent vectors $W_1, ...$ $,W_k$ and $W_{k+1},...,W_n$, respectively, which together form a basis for $\mathbf{R}^n$,

$$\phi : \mathbf{R}^k \to \mathbf{R}^n ,$$

$$\varphi : \mathbf{R}^{n-k} \to \mathbf{R}^n .$$

Here $\mathbf{t_k} = (t_1,...,t_k)$, $\mathbf{t_{n-k}} = (t_{k+1},...,t_n)$. Assume further that M^k and N^{n-k} both intersect $B_\delta(u_*)$, the ball of radius δ and center u_*. Then there exists a constant $C > 1$ such that if

$$\varepsilon < 1/(CM_0),$$

then M^k and N^{n-k} intersect each other at a unique point inside of the ball $B_\gamma(u_*)$, where

$$\gamma = CM_0{}^2\delta.$$

Here $M_0>1$ denotes a constant that relates the Euclidean norm on $\mathbf{R}^n$ to the norm determined by the basis $W_1,\ldots,W_n$ in the sense that

$$(1/M_0)|\alpha| < |\alpha_1 W_1 + \cdots \alpha_n W_n| < M_0|\alpha|$$

holds for all $\alpha \in \mathbf{R}^n$.

Now given U_L , and U_R , define

$$\phi(t_k) = T^R{}_{t_k} \circ T^{k-1}{}_{t_{k-1}} \circ \cdots \circ T^1{}_{t_1} U_L ,$$

$$\varphi(t_{n-k}) = T^{-(k+1)}{}_{t_{k-1}} \circ \cdots \circ T^{-n}{}_{t_n} U_R .$$

Here T^{-j} denotes the inverse of the function T^j . The existence and uniqueness of solutions of the Riemann problem then follows directly from the next lemma (details will appear in the authors' forthcoming paper.):

LEMMA 2: For U_L and U_R in a δ-neighborhood of U_* , ϕ and φ define Lipshitz continuous manifolds at level $a=a_R$, with ε-approximate tangent vectors $W_j = PR_j(U_*)$ for $j<k$, and $W_j = R_j(U_*)$ for $j \geq k$, where P denotes the matrix that projects onto the tangent space of $\mathcal{T}$ at $U=U_*$, and

$$\varepsilon = o(\delta) .$$

We note that it is the continuity of the curves $T^R(U_L)$ that leads to the existence and uniqueness of solutions of the Riemann problem, and to the fact that solutions of the Riemann problem depend continuously on the left and right states U_L and U_R. The Lipschitz continuity of the wave curves follows directly from the fact that the equations are posed in conservation form.

In conclusion, the general structure of the solutions in a neighborhood of U_* can be described as follows: to leading order the waves in the 0,k-characteristic families correspond to the waves in the Riemann problem solution for the scalar inhomogeneous equation; the general solution is obtained by adjoining to these waves the faster and slower waves from families $i \neq k$. Thus, the Riemann problem solutions of the scalar inhomogeneous equation give the canonical structure of solutions under our generic assumptions, just as the scalar homogeneous equation determines the local structure to leading order in the strictly hyperbolic case.

REFERENCES

[1] Courant and Freidrichs, *Supersonic flow and shock waves, John Wiley, New York, 1948.*

[2] H. Freisthuler, Rotational Degeneracy of Hyperbolic Systems of Conservation Laws, *Arch. Rat. Mech. Anal.,* to appear

[3] J.Glimm, Solutions in the large for nonlinear hyperbolic systems of equations, *Comm. Pure Appl. Math.* **18** (1965), 697-715

[4] E. Isaacson, Global solution of a Riemann problem for a non-strictly hyperbolic system of conservation laws arising in enhanced oil recovery, Rockefeller University preprint

[5] E. Isaacson, D. Marchesin, B. Plohr, and B. Temple, The Reimann

problem near a hyperbolic singularity: the classification of solutions of quadratic Riemann problems I, *SIAM J. Appl. Math.*, **48** No. 5 (1988)

[6] E. Isaacson and B. Temple, The structure of asymptotic states in a singular system of conservation laws, *Adv. Appl. Math.*, to appear

[7] E. Isaacson and B. Temple, Analysis of a singular system of conservation laws, *Jour. Diff. Equn.*, 65 No. 2 (1986)

[8] E. Isaacson and B. Temple, Examples and classification of non-strictly hyperbolic systems of conservation laws, *Abstracts of AMS*, January 1985.

[9] B. Keyfitz and H. Kranzer, A system of non-strictly hyperbolic conservation laws arising in elasticity theory, *Arch. Rat. Mech. Anal.* **72** (1980), 219-241

[10] P. D. Lax, Hyperbolic systems of conservation laws, II, *Comm. Pure Appl. Math.* **10** (1957), 537-566

[11] T. P. Liu, Quasilinear Hyperbolec Systems, *Comm. Math. Phys.*, **68**, 141-172 (1979)

[12] T.P. Liu, Scalar example

[13] D. Marchesin and P. J. Paes-Leme, PUC report.

[14] D. Schaeffer and M. Shearer, The classification of 2x2 systems of conservation laws, with application to oil recovery, with Appendix by D. Marchesin, P. J. Paes-Leme, D. Schaeffer, and M. Shearer, *Comm. Pure Appl. Math.*, **15** (1987), pp. 141-178.

[15] J.Smoller, *Shock Waves and Reaction Diffusion Equations*, Springer-Verlag, Berlin, Niw York, 1980

[16] B. Temple, Global solution of the Cauchy problem for a class of 2x2 nonstrictly hyperbolic conservation laws, *Adv. in Appl. Math.* **3** (1982), 335-375

Contemporary Mathematics
Volume **108**, 1990

BIFURCATION AND STABILITY PROBLEMS IN ROTATING PLANE COUETTE-POISEUILLE FLOW

George H. Knightly[1] & D. Sather

ABSTRACT. The mathematical problem of the flow of a viscous incompressible fluid between concentric, rotating, sliding cylinders with a pressure gradient applied is formulated, in its narrow gap limit, as an abstract problem in Hilbert space. The problem depends on Reynolds number and a parameter measuring, e.g., a swirl or spiral property of the flow. The bifurcation and stability of secondary states is investigated. In some cases the relative influence of pressure-driven (Poiseuille) effects versus boundary-driven (Couette) effects on stability is determined.

1. INTRODUCTION. Fluid flow between rotating cylinders provides a classical setting for experimental and theoretical investigation of fluid behavior. Here we present a unified approach to a class of such problems in the case of the so-called "narrow gap" where the mathematical structure is somewhat more manageable. We are particularly concerned with the relative importance of different driving forces for the bifurcation and stability of states of the system.

Let $\mathbf{u}$ denote the velocity vector, p the scalar pressure, of a viscous incompressible fluid occupying the annular region between rigid coaxial cylinders of radii a and $b > a$. Let (r, θ, x) denote cylindrical coordinates with x measured along the common axis $\mathbf{e}_x$ of the cylinders, so that the inner and outer cylinders correspond to $r = a$ and $r = b$, respectively. In a coordinate system that rotates about the x-axis with constant angular velocity Ω, the Navier-Stokes equations are

$$\frac{\partial \mathbf{u}}{\partial t} + 2\Omega \mathbf{e}_x \times \mathbf{u} + \mathbf{u} \cdot \nabla \mathbf{u} = -\nabla p + \nu \nabla^2 \mathbf{u}, \tag{1a}$$

$$\nabla \cdot \mathbf{u} = 0, \tag{1b}$$

1980 *Mathematics Subject Classification* (1985 *Revision*). Primary 76E30, 35Q10.
[1]This research was supported in part by ONR Grant N00014-88-K-0098.
This paper is in final form and no version of it will be submitted for publication elsewhere.

where ν denotes the (constant) kinematic viscosity. The inner and outer cylinders rotate about their axis, with angular rates Ω_a and Ω_b, respectively, and slide parallel to their axis at constant speeds T_a and T_b, respectively. In addition, a constant axial pressure gradient, $-\partial p/\partial x = P_x$, and a constant circumferential pressure gradient $-\partial p/\partial \theta = P_\theta$ are maintained. One obtains a basic steady flow $\mathbf{u} = \mathbf{U}(r)$ and investigates secondary states $\mathbf{u} = \mathbf{U} + \mathbf{v}$. When the relative gap $(b-a)/a$ is vanishingly small, one obtains for steady disturbances $\mathbf{v}$ a problem in rectangular coordinates that we shall reformulate as an equation

$$(*) \qquad 0 = \mathbf{v} - \lambda L\mathbf{v} - F(\mathbf{v}), \qquad \mathbf{v} \in \mathcal{H}, \qquad \lambda \in \mathbf{R}^1,$$

in a suitable Hilbert space $\mathcal{H}$. Here L is a linear operator, F is a nonlinear operator and λ is Reynolds number.

In general, the operator L depends upon one or more "structural" parameters corresponding to the various Couette (boundary-driven) and Poiseuille (pressure-driven) components of the basic flow and to the kinds of disturbances assumed. When the Couette and Poiseuille components are in constant proportion, measured by a parameter γ, and disturbances are sought in the direction determined by the basic flow, then L may be taken to depend on the single parameter γ. The operators involved are closely related to those appearing in [11] where purely circumferential flows were investigated. Here we can utilize much of the development in [11] to obtain corresponding results when both axial and circumferential forces are applied. We show for each γ in some interval, $|\gamma| < \gamma_0$, that there exists a nontrivial secondary state branching from the basic flow at a critical value $\lambda_c = \lambda_c(\gamma)$ of Reynolds number. Furthermore, we are able to show that $\lambda_c(\gamma) < \lambda_c(0)$. Thus, in general, the superposition of a small Poiseuille component on sliding and rotating Couette flow is destablizing. Such a result was proved in [11] for purely rotating Couette flow and conjectured in [13] for non-rotating flows on the basis of numerical calculations for the linear Navier-Stokes equations.

The outline of the paper is as follows. In Section 2 we present the basic flow and the narrow gap problem, utilizing the developments in Chandrasekhar [1], Drazin and Reid [4] and, especially, Joseph [6]. In Section 3 we introduce our Hilbert space setting for the general problem. Several methods have been developed to deal with problems such as $(*)$ in which additional structural parameters appear (e.g., see Chow, Hale, and Mallet-Paret [2, 3], the work [10;11;14] and the references cited in those papers). In Section 4 we follow the procedure in [11] in analyzing a particular case in which a Couette flow having both axial and circumferential components is perturbed by a Poiseuille flow when the spiral flow angles are constant.

2. THE CLASSICAL PROBLEM. We seek a basic steady flow

$$\mathbf{U} = U_r\mathbf{e}_r + U_\theta\mathbf{e}_\theta + U_x\mathbf{e}_x,$$

where $\mathbf{e}_s$ denotes a unit vector in the direction of coordinate s, subject to the stated forces and obeying the adherence condition: at the boundaries the velocity is that of the appropriate cylinder. Since the data have no components in the $\mathbf{e}_r$ direction, one takes $U_r = 0$ and looks for $\mathbf{U} = \mathbf{U}(r)$ independent of θ and x. Then $\nabla \cdot \mathbf{U} = 0$ holds identically and the system (1) becomes

$$-\frac{1}{r}U_\theta^2 - 2\Omega U_\theta = -\frac{\partial P}{\partial r} \equiv P_r, \tag{2a}$$

$$0 = \frac{1}{r}P_\theta + \nu(U_\theta'' + \frac{1}{r}U_\theta' - \frac{1}{r^2}U_\theta), \tag{2b}$$

$$0 = P_x + \nu(U_x'' + \frac{1}{r}U_x'), \tag{2c}$$

where the prime denotes differentiation with respect to r. The boundary conditions for (2) are

$$U_\theta(a) = a(\Omega_a - \Omega),\ U_\theta(b) = b(\Omega_b - \Omega),\ U_x(a) = T_a,\ U_x(b) = T_b. \tag{3}$$

Problem (2),(3) has the solution (e.g., see Joseph [6, p.160], Chandrasekhar [1, §77])

$$U_\theta(r) = \frac{P_\theta}{2\nu}(-r\log r + A_1 r + \frac{A_2}{r}) + A_3 r + \frac{A_4}{r}, \tag{4a}$$

$$U_x(r) = \frac{P_x}{4\nu}(-r^2 + A_5\log r + A_6) + A_7\log r + A_8, \tag{4b}$$

$$P = -(rP_r + \theta P_\theta + xP_x), \tag{4c}$$

where

$$A_1 = \frac{b^2\log b - a^2\log a}{b^2 - a^2},\quad A_2 = \frac{a^2b^2\log\frac{a}{b}}{b^2 - a^2},$$

$$A_3 = \frac{b^2(\Omega_b - \Omega) - a^2(\Omega_a - \Omega)}{b^2 - a^2},\quad A_4 = \frac{a^2b^2(\Omega_a - \Omega_b)}{b^2 - a^2},$$

$$A_5 = \frac{b^2 - a^2}{\log\frac{b}{a}},\quad A_6 = \frac{a^2\log b - b^2\log a}{\log\frac{b}{a}},$$

$$A_7 = \frac{T_b - T_a}{\log\frac{b}{a}},\quad A_8 = \frac{T_a\log b - T_b\log a}{\log\frac{b}{a}}.$$

The stability of the basic solution (4) may be studied by setting $\mathbf{u} = \mathbf{U} + \mathbf{v}$, $p = P + q$ in (1) and investigating the resulting problem for the disturbance quantities $\mathbf{v}, q$:

$$\frac{\partial \mathbf{v}}{\partial t} + 2\Omega\mathbf{e}_x \times \mathbf{v} + \mathbf{U}\cdot\nabla\mathbf{v} + \mathbf{v}\cdot\nabla\mathbf{U} + \mathbf{v}\cdot\nabla\mathbf{v} = -\nabla q + \nu\nabla^2\mathbf{v}, \tag{5a}$$

$$\nabla\cdot\mathbf{v} = 0, \tag{5b}$$

$$\mathbf{v} = \mathbf{0} \text{ at } r = a \text{ and } r = b. \tag{5c}$$

In the cylindrical coordinate system the operators in equation (5a) have the following meanings for arbitrary vectors $\mathbf{v} = v_r\mathbf{e}_r + v_\theta\mathbf{e}_\theta + v_x\mathbf{e}_x$ and $\mathbf{w} = w_r\mathbf{e}_r + w_\theta\mathbf{e}_\theta + w_x\mathbf{e}_x$ (e.g., see [1, p.292]):

$$\mathbf{e}_x \times \mathbf{w} = -w_\theta\mathbf{e}_r + w_r\mathbf{e}_\theta,$$

$$\mathbf{v}\cdot\nabla\mathbf{w} = v_r\frac{\partial\mathbf{w}}{\partial r} + \frac{1}{r}v_\theta\frac{\partial\mathbf{w}}{\partial\theta} + v_x\frac{\partial\mathbf{w}}{\partial x} - \frac{v_\theta w_\theta}{r}\mathbf{e}_r + \frac{v_\theta w_r}{r}\mathbf{e}_\theta,$$

$$\nabla^2 \mathbf{w} = \Delta \mathbf{w} - \frac{1}{r^2}[(w_r + 2\frac{\partial w_\theta}{\partial \theta})\mathbf{e}_r + (-2\frac{\partial w_r}{\partial \theta} + w_\theta)\mathbf{e}_\theta],$$

$$\Delta = \frac{\partial^2}{\partial r^2} + \frac{1}{r}\frac{\partial}{\partial r} + \frac{1}{r^2}\frac{\partial^2}{\partial \theta^2} + \frac{\partial^2}{\partial x^2}.$$

Here we shall treat, exclusively, the "narrow gap" situation in which $b - a$ is small relative to b. Thus, we set $\eta = a/b$ and introduce new dimensionless variables by

$$\tilde{x} = \frac{x}{b-a}, \ \tilde{y} = \frac{b\theta}{b-a}, \ \tilde{z} = \frac{r-a}{b-a}, \ \tilde{t} = \frac{\hat{U}t}{\lambda(b-a)}, \ \tilde{\mathbf{v}} = \frac{\lambda \mathbf{v}}{\hat{U}}, \ \tilde{q} = \frac{\lambda^2 q}{\hat{U}^2},$$

where $\hat{U} = [V_P^2 + V_C^2 + W_P^2 + W_C^2]^{\frac{1}{2}}$, $\lambda = (b-a)\hat{U}/\nu$ is Reynolds number and

(6) $$V_P = \frac{P_\theta(b-a)^2}{2\nu b}, V_C = b(\Omega_a - \Omega_b), W_P = \frac{P_x(b^2-a^2)^2}{2\nu b^2}, W_C = T_a - T_b.$$

Rewriting the basic solution (4) and the system (5) in the new variables, dropping the tilda's, taking $\Omega = \Omega_b$ and retaining only the leading terms in $(1-\eta)$, as $\eta \to 1$, one eventually obtains the problem (7) (below) for the narrow gap,

$$G = \{(x,y,z) : -\infty < x, y < \infty, 0 < z < 1\}.$$

(Henceforth, we denote by $\mathbf{v} = (v_1, v_2, v_3)$ the components of a vector $\mathbf{v}$ in the rectangular (x,y,z)-coordinates. Note that this entails a reversal of our previous, (r,θ,x), ordering of components.) Thus,

(7a) $$\frac{\partial \mathbf{v}}{\partial t} + \mathbf{v}\cdot\nabla\mathbf{v} + \lambda\mathcal{L}\mathbf{v} = -\nabla q + \Delta\mathbf{v}, \quad (x,y,z)\in G, \quad t \in \mathbf{R}^1,$$

(7b) $$\nabla\cdot\mathbf{v} = 0,$$

(7c) $$\mathbf{v} = 0 \text{ at } z = 0 \text{ and at } z = 1,$$

where ∇ and Δ have their usual meanings in rectangular coordinates,

(8) $$\mathcal{L}\mathbf{v} = S(v_3\mathbf{e}_y - v_2\mathbf{e}_z) + (V\frac{\partial}{\partial y} + W\frac{\partial}{\partial x})\mathbf{v} + v_3(W'\mathbf{e}_x + V'\mathbf{e}_y),$$

the prime now denotes d/dz, $S = 2\Omega(b-a)/\hat{U}$ is a swirl-like parameter and

(9a) $$V = \hat{U}^{-1}[V_P z(1-z) + V_C(1-z)],$$

(9b) $$W = \hat{U}^{-1}[W_P z(1-z) + W_C(1-z)].$$

The *basic spiral flow angle*, χ, is the angle the basic flow velocity in (4) makes with the direction $\mathbf{e}_x$. $\chi(r)$ depends only on r and is given by

$$\tan\chi(r) = U_\theta(r)/U_x(r).$$

In the narrow gap limit this becomes

(10) $$\tan\chi(z) = V(z)/W(z) = \frac{V_P z + V_C}{W_P z + W_C}.$$

Thus,

$$V(z) = \sqrt{V(z)^2 + W(z)^2}\,\sin\chi, \quad W(z) = \sqrt{V(z)^2 + W(z)^2}\,\cos\chi.$$

Experimental states are sometimes observed (e.g., see the photographs in [12] or in [6, pp. 176, 177]) that seem to possess special (spiral) directions in which the flow does not vary. One may look for such spiral disturbances theoretically by assuming, e.g., that the flow in (7) possesses a *disturbance spiral flow angle* $\psi = \psi(z)$ such that $\mathbf{v}, q$ are independent of the coordinate in the direction ψ.

The spiral flows are simpler when V and W are in fixed proportion so that $W(z) = \varepsilon V(z)$ for some constant ε. This is equivalent to $(W_P, W_C) = \varepsilon(V_P, V_C)$, important special cases of which are rotating plane Couette flow, where $V_P = W_P = 0$, $\varepsilon = W_C/V_C$, and rotating plane Poiseuille flow, where $V_C = W_C = 0$, $\varepsilon = W_P/V_P$. Then χ is independent of z and we look for spiral disturbances having constant flow angle ψ. We may introduce new coordinates (x', y', z) so that the disturbance is independent of x' and $\mathbf{e}_{x'}$ makes the angle ψ with $\mathbf{e}_x$. In these coordinates $\mathbf{e}_x = \cos\psi\mathbf{e}_{x'} - \sin\psi\mathbf{e}_{y'}$, $\mathbf{e}_y = \sin\psi\mathbf{e}_{x'} + \cos\psi\mathbf{e}_{y'}$ and

(11) $$\mathbf{U} = W\mathbf{e}_x + V\mathbf{e}_y = U_0(z)[\cos(\chi - \psi)\mathbf{e}_{x'} + \sin(\chi - \psi)\mathbf{e}_{y'}],$$

where

$$U_0(z) = (1 - z)\frac{zV_P + V_C}{\sqrt{V_P^2 + V_C^2}}.$$

Then the domain, $G_0' = \{(y', z) : -\infty < y' < \infty,\ 0 < z < 1\}$ is planar,

$$\nabla = \mathbf{e}_{y'}\frac{\partial}{\partial y'} + \mathbf{e}_z\frac{\partial}{\partial z}, \text{ and } \Delta = \frac{\partial^2}{\partial y'^2} + \frac{\partial^2}{\partial z^2}.$$

Throughout the remainder of this paper we consider only χ, ψ and S such that $0 < S\sin\psi < \hat{V}_C\cos(\chi - \psi)$. We scale the x'-component of $\mathbf{v}$ by

$$[(\hat{V}_C\cos(\chi - \psi) - S\sin\psi)/\sin\psi]^{\frac{1}{2}},$$

where $\hat{V}_C = V_C/\sqrt{V_C^2 + V_P^2}$, and we set $q = Q + \lambda S\tilde{q}\cos\psi$, where

$$\tilde{q} = \int^{(y',z)} (v_3 dy - v_2 dz).$$

Here $\tilde{q}$ is well defined and indpendent of path, since G_0' is simply connected and v is divergence free. We also define parameters

(12a) $$\gamma = \hat{V}_P\cos(\chi - \psi)/[\hat{V}_C\cos(\chi - \psi) - S\sin\psi], \quad \kappa_1 = V_P/V_C,$$

(12b) $$\kappa_2 = \hat{V}_C\sin(\chi - \psi)/\kappa_3, \quad \kappa_3 = [S\sin\psi(\hat{V}_C\cos(\chi - \psi) - S\sin\psi)]^{\frac{1}{2}}.$$

Making these substitutions and then omitting the primes we again obtain a problem of the form (7) but now with $\lambda = \kappa_3(b - a)\hat{U}/\nu$ and with operator $\mathcal{L}$ given by

(13) $$\mathcal{L}\mathbf{v} = -\mathcal{K}\mathbf{v} + \kappa_2(1 - z)(1 + \kappa_1 z)\frac{\partial\mathbf{v}}{\partial y},$$

where

$$\mathcal{K} = \begin{pmatrix} 0 & 0 & k_1 \\ 0 & 0 & \kappa_2 k_2 \\ 1 & 0 & 0 \end{pmatrix},$$

$k_1 = 1 - \gamma(1-2z)$ and $k_2 = 1 - \kappa_1(1-2z)$.

We shall require the solution of the linear eigenvalue problem on G_0

(14a) $$\lambda \mathcal{L}_0 \mathbf{v} = -\nabla Q + \Delta \mathbf{v},$$

(14b) $$\nabla \cdot \mathbf{v} = 0,$$

(14c) $$\mathbf{v} = \mathbf{0} \text{ at } z = 0, 1, \quad \mathbf{v} \text{ periodic in } y \text{ with period } 2\pi/\alpha,$$

where $\mathcal{L}_0$ is obtained from $\mathcal{L}$ by setting $\gamma = \kappa_2 = 0$ in (13). The problem (14) is well studied (see [1;4;5]). The eigenfunctions are complete and are obtained from the relations.

(15a) $$\mathbf{v} = e^{i\sigma y}(\phi_1(z), \phi_2(z), \phi_3(z)),$$

(15b) $$Q = e^{i\sigma y}\sigma^{-2} D^2 \phi_3'(z),$$

(15c) $$\phi_2 = i\sigma^{-1}\phi_3',$$

where $D^2 = \dfrac{d^2}{dz^2} - \sigma^2$, a prime denotes $\dfrac{d}{dz}$ and ϕ_1, ϕ_3 satisfy

(16a) $$D^4 \phi_3 - \mu\sigma^2 \phi_1 = 0,$$

(16b) $$D^2 \phi_1 + \mu\sigma^2 \phi_3 = 0,$$

(16c) $$\phi_1 = \phi_3' = \phi_3 = 0 \text{ at } z = 0, 1.$$

One can show for $\sigma > 0$ (e.g., see [5]) that the eigenvalue problem (16) has a countable number of positive simple eigenvalues, $0 < \mu_1(\sigma) < \mu_2(\sigma) < \cdots$, depending continuously on σ. Moreover, $\mu_1(\sigma) \to \infty$ as either $\sigma \to 0^+$ or $\sigma \to \infty$. Consequently, $\mu_1(\sigma)$ assumes an absolute minimum at some $\sigma_0 > 0$. *We assume that σ_0 is unique so that $\mu_1(\sigma) > \mu_1(\sigma_0)$ if $\sigma \neq \sigma_0$.* (This property is suggested by numerical calculations (e.g., see [1, §15(b)] and [4, §10]) and is usually assumed in such problems.) For some given integer $p_0 \geq 1$ we now choose α so that

(17) $$\sigma_0 = p_0 \alpha$$

and use this α to define the domain $G_\alpha = \{(y, z) : -\pi/\alpha < y < \pi/\alpha,\ 0 < z < 1\}$.

We may now determine a complete solution of the linear problem (14). Since $e^{i\sigma y}$ in (15) must have period $2\pi/\alpha$ in y, it follows that the only wave numbers, σ, corresponding to eigenfunctions having the required period in y are those for which $\sigma^2 = p^2\alpha^2$ for some integer p. For each p $(p = 1, 2, \cdots)$ the eigenvalue problem (16) has an infinite sequence of real, nontrivial solutions,

$$(\mu, \phi_1, \phi_3) = (\mu_{pq}, \phi_1^{pq}, \phi_3^{pq}) \qquad p = 1, 2, \cdots; \; q = \pm 1, \pm 2, \cdots.$$

Since $(-\mu, -\phi_1, \phi_3)$ is a solution of (16) whenever (μ, ϕ_1, ϕ_3) is a solution of (16), we may order the indices so that

$$\phi_1^{p(-q)} = -\phi_1^{pq}, \; \phi_3^{p(-q)} = \phi_3^{pq}, \; \mu_{p(-q)} = -\mu_{pq}, \text{ and } 0 < \mu_{p1} < \mu_{p2} < \cdots.$$

Using this notation, we see from our assumption (17) on α that

(18) $$\mu_0 \equiv \min_p \mu_{p1} = \mu_{p_0 1}$$

so that $\mu_{pq} > \mu_0$ if $(p, q) \neq (p_0, 1)$, $q \geq 1$.

The above discussion of the underlying problem (16) shows that the full eigenvalue problem (14) has the solutions

(19) $$\lambda = \mu_{pq} \text{ and } \mathbf{v} = \zeta^{mpq}, \qquad m = 0, 1; \; p = 1, 2, \cdots; \; q = \pm 1, \pm 2, \cdots,$$

where

(20) $$\zeta^{mpq} = e^{i(-1)^m p\alpha y}(\phi_1^{pq}, \phi_2^{pq}, \phi_3^{pq})$$

with Q determined by (15b) and ϕ_2^{pq} determined by (15c).

3. THE ABSTRACT PROBLEM. To investigate the bifurcation of stationary solutions of (7) from the trivial solution, we first recast the problem in a Hilbert space, in which we shall seek solutions having a doubly periodic structure. Thus, given positive numbers α_1 and α_2, we set

$$G = \{(x, y, z) : 0 < x < 2\pi/\alpha_1, 0 < y < 2\pi/\alpha_2, 0 < z < 1\}.$$

We define the complex Hilbert space, $\mathcal{H}$, as the closure of the set of smooth, divergence-free vector fields that vanish near $z = 0, 1$ and are periodic in x with period $2\pi/\alpha_1$, and periodic in y with period $2\pi/\alpha_2$, in the norm associated with the inner product

$$(\mathbf{v}, \mathbf{w}) = \int_G \sum_{j=1}^{3} \nabla v_j \cdot \nabla \bar{w}_j,$$

where the overbar denotes complex conjugation.

If we take the scalar product of (7a) with $\bar{\mathbf{w}} \in \mathcal{H}$, use (7b,c) and integration by parts, then for time-independent $\mathbf{v}$ we obtain

(21) $$(\mathbf{v}, \mathbf{w}) - \lambda(L\mathbf{v}, \mathbf{w}) = (F(\mathbf{v}), \mathbf{w}).$$

Here

$$F(\mathbf{v}) = \Phi(\mathbf{v}, \mathbf{v}), \qquad \mathbf{v} \in \mathcal{H},$$

and the linear operator $L : \mathcal{H} \to \mathcal{H}$ and bilinear operator $\Phi : \mathcal{H}\times\mathcal{H} \to \mathcal{H}$ are defined (weakly) by

(22)
$$(L\mathbf{v},\mathbf{w}) = -\int_G (\mathcal{L}\mathbf{v})\cdot\bar{\mathbf{w}},$$

(23)
$$(\Phi(\mathbf{u},\mathbf{v}),\mathbf{w}) = -\int_G (\mathbf{u}\cdot\nabla\mathbf{v})\cdot\bar{\mathbf{w}},$$

for all $\mathbf{u},\mathbf{v},\mathbf{w} \in \mathcal{H}$. Since in (21) $\mathbf{w}$ is an arbitrary element of $\mathcal{H}$, we see that a steady smooth solution of (7) in $\mathcal{H}$ satisfies the operator equation

(∗)
$$0 = \mathbf{v} - \lambda L\mathbf{v} - F(\mathbf{v}), \qquad \mathbf{v}\in\mathcal{H}, \qquad \lambda \in \mathbf{R}^1.$$

In fact, one can apply standard regularity methods (e.g.,see [7;8]) to show that the steady problems (7) and (∗) are equivalent.

In the case of spiral disturbances, when $\mathcal{L}$ is given by (13), one makes the obvious modifications for a planar domain $G_\alpha = \{(y,z) : -\pi/\alpha < y < \pi/\alpha, 0 < z < 1\}$ and again obtains (∗) with

(24)
$$L = L_0 - \gamma M_1 - \kappa_2 M_2 - \kappa_1\kappa_2 M_3,$$

where, for all $\mathbf{v},\mathbf{w} \in \mathcal{H}$,

(25)
$$(L_0\mathbf{v},\mathbf{w}) = -\int_{G_\alpha} (\mathcal{L}_0\mathbf{v})\cdot\bar{\mathbf{w}} = \int_{G_\alpha} (v_3\bar{w}_1 + v_1\bar{w}_3),$$

(26)
$$(M_1\mathbf{v},\mathbf{w}) = \int_{G_\alpha} (1-2z)v_3\bar{w}_1,$$

(27)
$$(M_2\mathbf{v},\mathbf{w}) = \int_{G_\alpha} [-v_3\bar{w}_2 + (1-z)\frac{\partial\mathbf{v}}{\partial y}\cdot\bar{\mathbf{w}}],$$

(28)
$$(M_3\mathbf{v},\mathbf{w}) = \int_{G_\alpha} [(1-2z)v_3\bar{w}_2 + (1-z)z\frac{\partial\mathbf{v}}{\partial y}\cdot\bar{\mathbf{w}}].$$

To investigate equation (∗) one requires further information on the operators L and F. In the general case one shows by standard methods (e.g., see [7]) that L is a bounded, linear, compact mapping from $\mathcal{H}$ to itself. For the special case in which L is given by (24) we will be particularly concerned with the eigenvalue problem when $\gamma = \kappa_2 = 0$, namely,

(29)
$$\mathbf{v} - \mu L_0\mathbf{v} = 0, \qquad \mathbf{v}\in\mathcal{H}, \qquad \mu\in\mathbf{R}^1.$$

The linear problem (29) is equivalent to the classical problem (14) and we have the following result.

LEMMA 1. *(i) The linear operator L_0 is selfadjoint and compact. Its characteristic values and eigenfunctions are given by (19). The eigenfunctions satisfy*

(30)
$$(\zeta^{mpq},\zeta^{rst}) = \delta_{mr}\delta_{ps}\delta_{qt},\ m,r=0,1; p,s=1,2,\cdots; q,t=\pm1,\pm2,\cdots,$$

where δ_{ij} is the usual Kronecker delta symbol. If μ_0 denotes the smallest positive characteristic value of L_0, then the nullspace $\mathcal{N} = \mathcal{N}(I-\mu_0 L_0)$ has dimension $n = 2$. (ii) The linear *operators $M_j : \mathcal{H} \to \mathcal{H}$, $j = 1,2,3$, are compact and satisfy $M_j : \mathcal{N} \to \mathcal{N}^\perp$, $j = 1,2$, where $\mathcal{N}^\perp$ denotes the orthogonal complement of $\mathcal{N}$ in $\mathcal{H}$. The adjoint operators M_j^*, $j = 1,2,3$, are characterized by*

$$(M_1^*\mathbf{v}, \mathbf{w}) = \int_{G_\alpha} (1-2z)v_1\bar{w}_3, \tag{31}$$

$$(M_2^*\mathbf{v}, \mathbf{w}) = \int_{G_\alpha} [-v_2\bar{w}_3 + (1-z)\mathbf{v}\cdot\frac{\partial\bar{\mathbf{w}}}{\partial y}], \tag{32}$$

$$(M_3^*\mathbf{v}, \mathbf{w}) = \int_{G_\alpha} [(1-2z)v_2\bar{w}_3 + (1-z)z\mathbf{v}\cdot\frac{\partial\bar{\mathbf{w}}}{\partial y}]. \tag{33}$$

(iii) The nonlinear operator $F(\mathbf{v}) = \Phi(\mathbf{v},\mathbf{v})$ is generated by the bounded bilinear operator Φ in (23), which satisfies $\Phi : \mathcal{H}\times\mathcal{H} \to \mathcal{H}$ and $\Phi : \mathcal{H}\times\mathcal{N} \to \mathcal{N}^\perp$.

Proof. The compactness properties in (i) and (ii) are essentially well-known (e.g., see[7,8]). The statements in (i) about characteristic values and eigenfunctions follow immediately from the equivalence of problems (29) and (14) and the development of problem (14) in Section 2; equation (30) is obtained as in [9, Appendix] after rescaling the eigenfunctions, ζ^{mpq}, by constants depending upon p and q. Since μ_0, defined in (18), is a simple eigenvalue of (16) and $\mu_0 < \mu_{pq}$ for $(p,q) \neq (p_0, 1)$, $q \geq 1$, μ_0 is also the smallest positive characteristic value of L_0. The associated nullspace, $\mathcal{N}$, of $I - \mu_0 L_0$ is spanned by ζ_0^0, ζ_0^1, where $\zeta_0^m \equiv \zeta^{mp_01}$, and $\mathcal{N}^\perp$ is spanned by $\{\zeta^{mpq} : (m,p,q) \neq (0,p_0,1),(1,p_0,1)\}$. Thus, (i) is proved. The formulas (31) - (33) follow easily from the definitions (26) - (28) of the M_j. The property $M_1 : \mathcal{N} \to \mathcal{N}^\perp$ is obtained as in part (ii) of Lemma 3.1 in [9] while the corresponding property for M_2 is a consequence of

$$(M_2\zeta_0^m, \zeta_0^r) = \delta_{mr}\int_{G_\alpha}\{\frac{\partial}{\partial y}[\frac{(1-z)}{2} \mid \zeta_0^m \mid^2] + \frac{\partial}{\partial z}[\frac{i}{2\sigma_0} \mid \zeta_{03}^m \mid^2]\} = 0 \tag{34}$$

which follows from (15a,c), the divergence theorem, (16c) and the periodicity of ζ_0^m in y. Finally, (iii) is proved in a manner similar to the proof of part (iv) of Lemma 3.1 in [9].

4. PERTURBATION OF COUETTE FLOW BY POISEUILLE FLOW. In this section we study problem $(*)$ for the special case of spiral disturbances when L is given by equation (24) with $\kappa_2 = 0$. Thus, we treat henceforth the case in which the basic flow has constant spiral flow angle χ and we look for spiral disturbances having the same spiral flow angle $\psi = \chi$. Then $L = L_0 - \gamma M_1$ with $\gamma = \hat{V}_P/[\hat{V}_C - S\sin\chi]$ and our assumptions on χ and S become $0 < S\sin\chi < \hat{V}_C$. We impose the further simplification of restricting equation $(*)$ for this problem to the real subspace $\mathcal{H}_0$ of $\mathcal{H}$ defined by $\mathcal{H}_0 = \{\mathbf{v} \in \mathcal{H} : \mathbf{v}$ has real components (v_1, v_2, v_3) with v_1 and v_3 even in y, v_2 odd in $y\}$. This has the effect of replacing each pair ζ^{0pq}, ζ^{1pq} of eigenfunctions of L_0 in (19) by the single eigenfunction

$$\psi^{pq} = \frac{1}{2}(\zeta^{0pq} + \zeta^{1pq}) = (\phi_1^{pq}\cos p\alpha y, -\sigma^{-1}\frac{d\phi_3^{pq}}{dz}\sin p\alpha y, \phi_3^{pq}\cos p\alpha y). \tag{35}$$

(Of course one could repeat this for $\mathcal{H}_0$ defined with v_1, v_3 odd in y, v_2 even in y and $\cos p\alpha y$ interchanged with $\sin p\alpha y$, but the solutions obtained this way are just translations in y of those obtained here). As a consequence of (i) of Lemma 1, the eigenfunctions ψ^{pq} in (35) are complete in $\mathcal{H}_0$ and satisfy

$$(\psi^{pq}, \psi^{st}) = \delta_{ps}\delta_{qt}. \tag{36}$$

The smallest characteristic value, μ_0, of L_0 is simple in $\mathcal{H}_0$ and corresponds to the eigenfunction $\psi_0 = \psi^{p_0 1}$, which spans the nullspace, $\mathcal{N}_0$, of $I - \mu_0 L_0$ on $\mathcal{H}_0 = \mathcal{N}_0 \oplus \mathcal{N}_0^{\perp}$. We collect these observations as part (i) of the following lemma; part (ii) can be derived as in Lemma 3.2 of [9].

LEMMA 2. (i.) *When equation* $(*)$ *and the operators* L_0, M_1, Φ *and* F *are restricted to* $\mathcal{H}_0$*, the properties of these operators given in Lemma 1 continue to hold with eigenfunctions of* L_0 *now given by (35), and nullspace* $\mathcal{N}_0$ *now having dimension* $n = 1$.

(ii) *The critical characteristic value,* $\lambda_c = \lambda_c(\gamma)$*, of* $L_\gamma \equiv L_0 - \gamma M_1$*, i.e., the positive characteristic value of* L_γ *of least magnitude, has the form*

$$\lambda_c(\gamma) = \mu_0 - \mu_0^3 b\gamma^2 + \Lambda(\gamma), \quad |\, \gamma \,| < \gamma_1, \tag{37}$$

where Λ *is real, analytic and satisfies* $\Lambda(\gamma) = O(\gamma^3)$ *as* $\gamma \to 0$. *Here* $b = (M_1 K M_1 \psi_0, \psi_0)$*, where* $K : \mathcal{N}_0^{\perp} \to \mathcal{N}_0^{\perp}$ *denotes the inverse of the restriction of* $I - \mu_0 L_0$ *to* $\mathcal{N}_0^{\perp}$.

Under the conditions of this section, problem $(*)$ becomes

$$0 = \mathbf{v} - \lambda L_\gamma \mathbf{v} - F(\mathbf{v}), \qquad \mathbf{v} \in \mathcal{H}_0, \quad \lambda \in \mathbf{R}^1, \quad \gamma \in \mathbf{R}^1, \tag{†}$$

and is now identical to the problem (†) in [11] obtained for *the case in which the cylinders do not slide axially and there is no axial Poiseuille component in the basic flow*. As a consequence, we are able to utilize the results in [11], as we now describe.

One seeks a solution of (†) in the form

$$\mathbf{v} = \gamma(\beta\psi_0 + \gamma\Psi), \qquad \lambda = \mu_0 - \mu_0\gamma^2(\mu_0 b - \tau), \tag{38}$$

where $\beta \in \mathbf{R}^1$, $\Psi \in \mathcal{N}_0^{\perp}$ and $\tau \in \mathbf{R}^1$ are to be determined. It is shown in [11] that given $\rho > 0$ there is a $\gamma_0 > 0$ such that if $|\, \gamma \,| < \gamma_0$ then the part of (†) in $\mathcal{N}_0^{\perp}$ has a unique solution $\Psi = \Psi(\beta, \tau, \gamma)$ for every $(\beta, \tau) \in \mathbf{R}^2$ with $\beta^2 + \tau^2 < \rho^2$. In fact, Ψ is analytic and of the form

$$\Psi = -\beta\mu_0 K M_1 \psi_0 + \beta^2 K F(\psi_0) + \gamma\Psi_1, \tag{39}$$

where $K = [(I - \mu_0 L_0)\,|_{\mathcal{N}_0^{\perp}}]^{-1}$ and $\Psi_1 = \Psi_1(\beta, \tau, \gamma) \in \mathcal{N}_0^{\perp}$ is bounded with bound depending only on ρ. The bifurcation equation is then found to be

$$0 = -\tau\beta + a\beta^2 + c\beta^3 + r(\beta, \tau, \gamma), \tag{40}$$

where

$$a \equiv \mu_0(M_1 K F(\psi_0), \psi_0) + \mu_0(\Phi(\psi_0, K M_1 \psi_0), \psi_0) + \mu_0(\Phi(K M_1 \psi_0, \psi_0), \psi_0),$$

$$c \equiv -(\Phi(\psi_0, KF(\psi_0)), \psi_0) - (\Phi(KF(\psi_0), \psi_0), \psi_0),$$

and for $\beta^2 + \tau^2 < \rho^2$, $|\gamma| < \gamma_0$ the remainder r is given by

$$\begin{aligned} r(\beta, \tau, \gamma) = &\gamma(\mu_0 M_1 \Psi_1 - \Phi(\beta\psi_0, \Psi_1) - \Phi(\Psi_1, \beta\psi_0) \\ &- F(\Psi) + \gamma\mu_0(\tau - \mu_0^2 b) M_1 \Psi, \psi_0). \end{aligned}$$

Moreover, since Φ satisfies the additional condition

$$(\Phi(u, v), w) = -(\Phi(u, w), v), \qquad u, v, w \in \mathcal{H}_0,$$

one can, as in [11], make use of parts *(v)* and *(vi)* of Lemma 3.1 in [9] to show in (40) that $a = 0$ always and $c > 0$ in essentially all cases. In addition, one can use the invariance of the system (†) under the translation $y \to y + \pi/\alpha$ to show that, for each τ and γ, the remainder term r in (40) is odd in β (see also Appendix B in [14] and the proof of Lemma 3.2. in [10]). Thus, r has the form $r(\beta, \tau, \gamma) = \beta s(\beta, \tau, \gamma)$, where s is analytic and satisfies

$$|s(\beta, \tau, \gamma)| \leq s_0 |\gamma|, \qquad \beta^2 + \tau^2 < \rho^2, \quad |\gamma| < \gamma_0,$$

with s_0 depending only on ρ. In light of these observations, we may replace the bifurcation equation (40) by

(41) $$0 = -\tau + c\beta^2 + s(\beta, \tau, \gamma) \equiv E(\beta, \tau, \gamma).$$

Since $E(\beta, \tau, 0) = 0$ on the part of the parabola $\tau = \beta^2$ inside the disc $\beta^2 + \tau^2 < \rho^2$ and since $\partial E/\partial\tau(\beta, \tau, 0) = -1$, one may invoke the implicit function theorem (as in the proof of Theorem 3.1 in [11]) to show that equation (41) has a solution.

The results of [11,§3] now apply directly. Combining Theorem 3.1, Remark 3.1 and Corollary 3.1 of [11] in the current context, we obtain the following result under the generic assumption that $c > 0$. Stability here means *linearized stability* in the space $\mathcal{H}_0$.

THEOREM 1. *(i). Given $\rho > 0$ there exists $\gamma_0 > 0$ such that for $|\gamma| < \gamma_0$ equation (41) has a solution*

(42) $$\tau = \tau(\beta, \gamma) \equiv \tau_0(\gamma) + \beta^2[c + \tau_1(\beta, \gamma)]$$

that is bounded, analytic and unique in

(43) $$B = \{(\beta, \tau, \gamma) : |\tau - c\beta^2| < k|\gamma|, \quad |\beta| < \rho, \quad |\gamma| < \gamma_0\}$$

where the constant k depends only on ρ, and the terms τ_0 and τ_1 are $O(\gamma)$ as $\gamma \to 0$, uniformly for $|\beta| < \rho$. For each fixed γ satisfying $|\gamma| < \gamma_0$, the corresponding nontrivial solution branch $(\mathbf{v}^*(\beta), \lambda^*(\beta))$ of (†), $|\beta| < \rho$, has the form (38) with Ψ and τ given by (39) and (42), respectively, and is stable for $|\beta| < \rho$.

Remark 1. (i) Theorem 1 shows, for each fixed γ satisfying $0 < |\gamma| < \gamma_0$ with γ_0 sufficiently small, that in the narrow gap limit the basic rotating, axial, circumferential Couette-Poiseuille flow is, for $\lambda < \lambda_c(\gamma)$, stable to disturbances having the same spiral flow angle as the basic flow. At $\lambda = \lambda_c(\gamma)$, however, the basic flow loses stability to a supercritical, stable roll-type solution of the form (38).

(ii) When $b > 0$ in (37), we have, for γ sufficiently small, $\lambda_c(\gamma) < \lambda_c(0) = \mu_0$, where we recall that μ_0 is the critical eigenvalue of the linearized problem for the rotating, axial, circumferential Couette flow (in the narrow gap limit). Thus, for a given value of the swirl S, $0 < S \sin\chi < \hat{V}_C$, the addition of a suitable, sufficiently-small component of Poiseuille flow to the Couette flow always leads to roll-type solutions at values of λ that are greater than $\lambda_c(\gamma)$ but less than the critical eigenvalue μ_0 at which the Couette flow loses stability. This property was proved in [11] when the basic flow has no axial component; earlier Potter [13] conjectured, on the basis of numerical calculations, this type of result for non-rotating Couette-Poiseuille flow.

(iii) That $b > 0$ does hold follows from the numerical result in [1, §71(d)] (see also [4, p.98]) for the narrow gap Taylor problem with the parameter $(1-\mu)/(1+\mu)$ in [1] replaced by our parameter γ.

References

[1] S. Chandrasekhar, Hydrodynamic and Hydromagnetic Stability, Clarendon Press, Oxford, 1961.

[2] S. Chow, J.K. Hale, & J. Mallet-Paret, "Applications of generic bifurcation. I.", Arch. Rational Mech. Anal. 59 (1975), 159-188.

[3] S. Chow, J.K. Hale, & J. Mallet-Paret, "Applications of generic bifurcation. II.", Arch. Rational Mech. Anal. 62 (1975), 209-235.

[4] P.G. Drazin & W.H. Reid, Hydrodynamic Stability, Cambridge University Press, Cambridge, 1981.

[5] V.I. Iudovich, "On the origin of convection", J. Appl. Math. Mech 30 (1966), 1193-1199.

[6] D.D. Joseph, Stability of Fluid Motions I, Springer Tracts in Natural Philosophy 28, Springer Verlag, New York, 1976.

[7] K. Kirchgässner, "Bifurcation in nonlinear hydrodynamic stability", SIAM Review 17 (1975), 652-683.

[8] K. Kirchgässner & H. Kielhöfer, "Stability and bifurcation in fluid dynamics", Rocky Mountain J. Math. 3 (1973), 275-318.

[9] G.H. Knightly & D. Sather, "A selection principle for Bénard-type convection", Arch. Rational Mech. Anal. 88 (1985), 163-193.

[10] G.H. Knightly & D. Sather, "Stability of cellular convection", Arch. Rational Mech. Anal. 97 (1987), 271-297.

[11] G.H. Knightly & D. Sather, "Structure parameters in rotating Couette-Poiseuille channel flow," Rocky Mt. J. Math. 18 (1988), 339-355.

[12] H.M. Nagib, "On instabilities and secondary motions in swirling flows through annuli, Ph.D. dissertation, Illinois Institute of Technology, 1972.

[13] M.C. Potter, "Stability of plane Couette-Poiseuille flow", J. Fluid Mech. 24 (1966) 609-619.

[14] D. Sather, "Primary and secondary steady flows of the Taylor problem", J. Differential Equations 71 (1988), 145-184.

DEPARTMENT OF MATHEMATICS
UNIVERSITY OF MASSACHUSETTS
AMHERST, MASSACHUSETTS 01002

&

DEPARTMENT OF MATHEMATICS
UNIVERSITY OF COLORADO
BOULDER, COLORADO 80309

Contemporary Mathematics
Volume **108**, 1990

Bifurcations of Central Configurations in the N-Body Problem

Kenneth R. Meyer and Dieter S. Schmidt†

ABSTRACT. This paper discusses a series of studies done by the authors on the bifurcations of central configurations in the N-body problem. Modern bifurcation analysis and algebraic processors like the general purpose processor MACSYMA and the special purpose processor POLYPAK were used to find a multitude of different bifurcations.

I Introduction. The study of central configurations (c.c.) of the N-body problem has had a long history starting with the famous collinear configuration of the 3-body problem found by Euler (1767). Over the intervening years many different technologies have been applied to the study of c.c. In the older papers of Euler (1767), Lagrange (1772), Hoppe (1879), Lehmann-Filhes (1891), Moulton (1910) et al. special coordinates, symmetries and analytic techniques were used. Dziobek (1900) used the theory of determinants; Smale (1970) used Morse theory; Palmore (1975) used homology theory; Simo (1977) used a computer; and Moeckel (1986) used real algebraic geometry in their investigations. Thus, the study of c.c. has been a testing ground for many different methodologies of mathematics.

In a series of papers, Meyer(1987), Meyer and Schmidt (1988a,1988b), Schmidt(1988), we have used the methods of modern bifurcation analysis and automated algebraic processor to study this subject. Specifically, in the first paper, Meyer(1987), a fold catastrophe or saddle-node type bifurcation was established in the four body problem by continuing a bifurcation in the restricted four body problem into the full four body problem with one small mass. The point of bifurcation was found using numerical methods and then established by rigorous analysis. In the second and third papers, Meyer and Schmidt (1988a,1988b), the bifurcations of a central configuration which consists of N-1 particles of mass 1 at the vertices of a regular polygon

1980 *Mathematics Subject Classification* (1985 *Revision*). 70F10.
This paper is in final form and no version of it will be submitted for publication elsewhere.
†This research was supported by a grant from the Applied and Computational Mathematics Program of DARPA administered by NIST.

and one particle of mass m at the centroid was studied. We call this the regular polygon central configuration (r.p.c.c.). In the second paper, we considered the 4 and 5 body problems and use the mutual distances as special coordinates following the lead of Dziobek(1900). These coordinates make the 4-body problem relatively easy to handle and the 5-body problem accessible, but beyond 5, Dziobek's coordinates become very cumbersome. The 4 and 5-body problem in these special coordinates are sufficiently simple that the general purpose algebraic processor MACSYMA could handle the tedious calculations. In the third paper the investigation of the bifurcations of these c.c. for larger N required the special purpose algebraic processor POLYPAK written by the second author because the computations increased rapidly with N. In the analysis of the 4 and 5 body problems the classical power series methods of bifurcation analysis handles the problem nicely, but for larger N a systematic use of Lie transforms by Deprit (1969) was mandated in order to bring the equations into a normal form. The first three papers dealt with the planar N-body problem, but in Schmidt(1988) the Dziobek coordinates were used with the aid of MACSYMA to find a bifurcation from a tetrahedron configuration of the spatial 5-body problem.

The problem of finding a c.c. can be reduced to finding a critical point of the potential energy function on the manifold of constant moment of inertia. Thus the problem falls within the domain of catastrophe theory and so the general theory is well understood. However, this specific problem has a high degree of symmetry, many variables and a constraint, so the computations must be performed with care. We consider these papers as case studies in bifurcation analysis in face of these complexities.

Even though as solutions of the N-body problem c.c. are quite rare and rather special, they are of central importance in the analysis of the asymptotic behavior of the universe. In general, solutions which expand beyond bounds or collapse in a collision do so asymptotically to a central configuration. A survey and entrance to this literature can be found in Saari (1980).

II Central Configurations for the N-body. The N-body problem is the system of differential equations which describe the motion of N particles moving under the influence of their mutual gravitational attraction. Let $q_j \in \mathbb{R}^3$ be the position vector, $p_j \in \mathbb{R}^3$ the momentum vector and $m_j > 0$ the mass of the jth particle, $1 \leq j \leq N$, then the equations of motion are

$$\dot{q}_j = \frac{\partial H}{\partial p_j} = \frac{1}{m_j} p_j,$$

(1)

$$\dot{p}_j = -\frac{\partial H}{\partial q_j} = \frac{\partial U}{\partial q_j}$$

where H is the Hamiltonian

$$H = \sum_{j=1}^{N} \frac{\| p_j \|^2}{2m_j} - U(q) \tag{2}$$

and U is the (self) potential

$$U = \sum_{1 \le i < j \le N} \frac{m_i m_j}{\| q_i - q_j \|} . \tag{3}$$

These equations reduce to the Newtonian formulation

$$m_j \ddot{q}_j = \frac{\partial U}{\partial q_j}, \quad j = 1,\ldots,N. \tag{4}$$

We seek a homothetic solution by setting $q_j = \phi(t)u_j$ where ϕ is a scalar function and the u_j are constants. Such a solution exists provided there is a constant λ so that the following equations are satisfied.

$$\ddot{\phi} = \frac{-\lambda}{\phi^2} \tag{5}$$

$$-\lambda m_j u_j = \frac{\partial U}{\partial u_j}, \quad j = 1,\ldots,N. \tag{6}$$

Equation (5) is just the differential equation of the collinear Kepler problem and so has many solutions. Equation (6) is an algebraic equation for the N vectors $u_1,\ldots,u_N$ and the scalar λ. If $u_1,\ldots,u_N$ satisfy (6) for some λ then $u_1,\ldots,u_N$ is called a *central configuration.*

It is classical and easy to verify that if $\bar{u} = (\bar{u}_1,\ldots,\bar{u}_N)$ and $\bar{\lambda}$ is a solution of (6) then the center of mass of $\bar{u}$ is at the origin ($\sum m_j \bar{u}_j = 0$) and $\bar{\lambda} = U(\bar{u})/2I(\bar{u}) > 0$ where I is the moment of inertia

$$I(u) = \frac{1}{2} \sum_{j=1}^{N} m_j \| u_j \|^2. \tag{7}$$

We will set

$$M = \{ u \in \mathbb{R}^{3N} : \sum m_j u_j = 0 \}$$

(8) $$\Delta = \{ u \in \mathbb{R}^{3N} : u_i = u_j \text{ for some } i \neq j \}$$

$$S = \{ u \in M : I(u) = 1 \}.$$

The variable λ can be considered as Lagrange multiplier and so an equivalent definition of a central configuration is a critical point of U restricted to $S \setminus \Delta$. If u is a c.c. then so is $Au = (Au_1, \ldots, Au_N)$ where $A \in O(3,\mathbb{R})$ is an orthogonal matrix. We can define an equivalence relation by $u \sim Au$ when $A \in O(3,\mathbb{R})$ and since U, I, are constant on equivalence classes we can define the quotient spaces $\mathcal{S} = (S \setminus \Delta) / \sim$ and the function $\mathcal{U} : \mathcal{S} \to R$ by $\mathcal{U}([u]) = U(u)$ where [ ] denotes an equivalence class. $\mathcal{S}$ and $\mathcal{U}$ are smooth. Thus a similarity class of c.c. is a critical point of $\mathcal{U}$.

A central configuration is called non-degenerate if its equivalence class is a non-degenerate critical point of $\mathcal{U}$ in the sense of Morse theory, i.e. the Hessian is non-singular at the critical point. It follows from the implicit function theorem that bifurcations can occur only at degenerate critical points, so the first quest is to find degenerate c.c.

III. The Restricted Problem. Consider the planar (N+1)-body problem with one particle small, so let $m_{N+1} = \varepsilon$ and $x = u_{N+1}$. The equations for a c.c. become

(1) $$-\lambda m_j u_j = \frac{\partial U_N}{\partial u_j} + O(\varepsilon), \quad j = 1,\ldots,N$$
$$-\lambda x = \frac{\partial W}{\partial x},$$

where $W = \sum_{i=1}^{N} (m_i/\|x-u_i\|)$ and U_N is the self potential of the N-body problem. When $\varepsilon = 0$ the equations in (1) decouple and a solution is an (N+1)-tuple $(\bar{u}_1,\ldots,\bar{u}_N,\bar{x})$ where $\bar{u}_1,\ldots,\bar{u}_N$ is a c.c. for the N-body problem and $\bar{x}$ is a critical point of

(2) $$V(x) = \sum_{i=1}^{N} \frac{m_i}{\| x - \bar{u}_i \|} + \frac{\lambda}{2} \|x\|^2.$$

Note that λ is determined by the fact that $\bar{u}_1,\ldots,\bar{u}_N$ is a c.c. The function V is called the potential of the restricted N-body problem.

It is not too hard to verify that if $\bar{u}_1,\ldots,\bar{u}_N$ is a nondegenerate c.c. and $\bar{x}$ is a nondegenerate critical point of (2) then for small ε the full (N+1)-body problem has a nondegenerate c.c close to $(\bar{u}_1,\ldots,\bar{u}_N,\bar{x})$. In Meyer(1987) it was proven that there is a degenerate c.c. in the full 4-body problem by showing that the restricted 4-body problem has a fold catastrophe and this fold catastrophe can be continued into the full 4-body problem.

Consider the one parameter family of the restricted 4-body problems where the c.c. of the 3-body problem is the equilateral triangle c.c. with

$$(3)\qquad \begin{aligned} m_1 &= 1 - \mu, & \bar{u}_1 &= (1, -\sqrt{3}\mu), \\ m_2 &= 1 - \mu, & \bar{u}_2 &= (-1, -\sqrt{3}\mu), \\ m_3 &= 2\mu, & \bar{u}_3 &= (0, \sqrt{3}(1-\mu)). \end{aligned}$$

For $\mu \simeq 0.4234$ this potential has a fold catastrophe, i.e. a critical point with $V_1 = V_2 = V_{12} = V_{22} = 0$ and $V_{11} \neq 0$, $V_{222} \neq 0$ and $V_{23} \neq 0$ where the subscripts 1,2,3 denote differentiation with respect to x_1, x_2 and μ respectively. Figure 1 shows the potential for $\mu = 0.2$; note the two minima and the saddle point in front of the middle pole. Figure 2 show the potential for $\mu = 0.5$; note that there is only one minimum in front of the middle pole and that one minimum and the saddle point are gone. The viewer for the three dimensional plots in Figures 1 and 2 is situated above the negative x_2-axis at a point that is on a line that makes a 60^0 with the V-axis. At the fold, the saddle point and one minimum come together and eliminate each other. A careful application of the implicit function theorem shows that this fold catastrophe persists in the full 4-body problem for ε small.

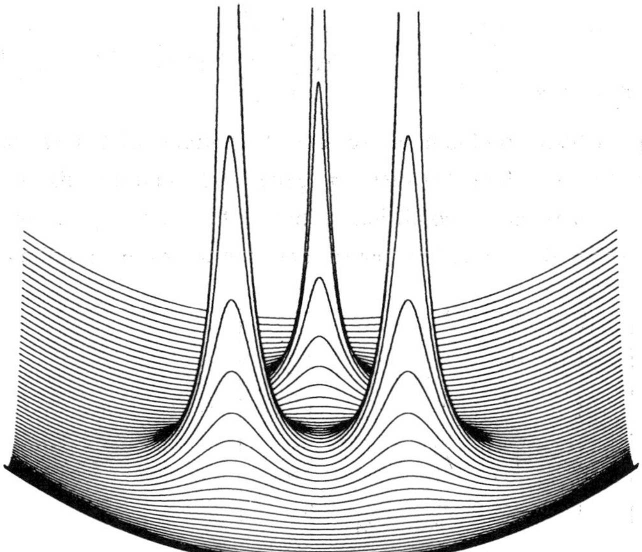

Figure 1. The potential V for $\mu = 0.2$.

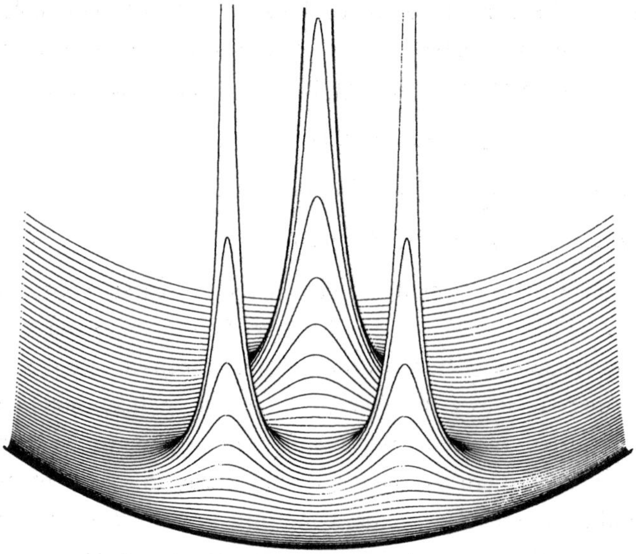

Figure 2. The potential V for $\mu = 0.5$.

IV. Dziobek Coordinate. Dziobek(1900) used the mutual distances $r_{ij} = \| u_i - u_j \|$ as coordinates in his study of c.c. in the planar 4-body problem. The potential, U, and the moment of inertia, I, can easily be expressed in terms of the mutual distances, in fact

$$(1) \qquad U = \sum_{1 \leq i < j \leq N} \frac{m_i m_j}{r_{ij}}, \quad I = \frac{1}{4M} \sum_{i=1}^{N} \sum_{j=1}^{N} m_i m_j r_{ij}^2, \quad M = \sum_{i=1}^{N} m_i.$$

For the planar 4-body problem there are 6 mutual distances and clearly they over determine the problem because in general 5 mutual distances suffice. A necessary and sufficient condition that the six positive numbers r_{ij}, $1 \leq i < j \leq N$ be the mutual distances between 4 collinear points is

$$(2) \qquad F = \begin{vmatrix} 0 & 1 & 1 & 1 & 1 \\ 1 & 0 & r_{12}^2 & r_{13}^2 & r_{14}^2 \\ 1 & r_{12}^2 & 0 & r_{23}^2 & r_{24}^2 \\ 1 & r_{13}^2 & r_{23}^2 & 0 & r_{34}^2 \\ 1 & r_{12}^2 & r_{24}^2 & r_{34}^2 & 0 \end{vmatrix} = 0.$$

This 19th century determinant is $288V^2$ where V is the volume of the tetrahedron whose 6 edges are given. $F = 0$ is simply another constraint.

Thus we follow Dziobek to find c.c. by studying the critical points of

(3) $W = U + \lambda I + \upsilon F$,

where λ and υ are Lagrange multipliers.

Consider the one parameter family of c.c. in the 4-body problem where 3 of the particles have mass 1 and are located at the vertices of an equilateral triangle and the fourth particles has mass m and is located at the centroid of the triangle. Palmore(1973) showed that for $m = m^* = (64\sqrt{3}+81)/249$ this c.c. is degenerate. We verify Palmore's result in Meyer and Schmidt(1988a) and also show that m^* is a point of bifurcation. For $m < m^*$ there is an acute isosceles triangle c.c. which approaches the equilateral family as $m \to m^*-$, see Figure 3a. For $m > m^*$ there is an obtuse isosceles triangle c.c. which approaches the equilateral family as $m \to m^*+$, see Figure 3b.

The computations are fairly lengthy and so MACSYMA was used to carry out the details. This method was also used in Meyer and Schmidt(1988a) to find a bifurcation in the planar 5-body problem which is similar to the bifurcation discussed above.

The Dziobek coordinates where used in Schmidt(1988) to find a bifurcation of the spatial 5-body problem. Consider a c.c. where 4 particles of mass 1 are placed at the vertices of a regular tetrahedron and one particle of mass m is placed at the centroid. For $m = m^{\#} = (10368+1701\sqrt{6})/54952$ this c.c. is degenerate and is also a point of bifurcation. A family of tetrahedron bifurcates from the regular tetrahedron in a manner similar to the bifurcation of the triangles in the planar problem.

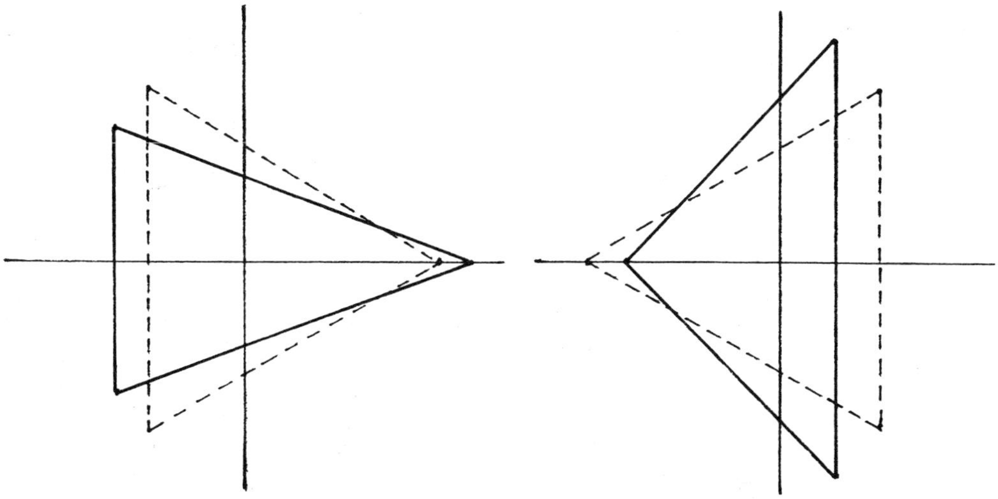

3a. $m < m^*$ 3b. $m > m^*$

Figure 3.

V. Large N. In Meyer and Schmidt(1988b) the bifurcations of the regular polygon central configuration for large N was investigated using different methods and a different algebraic processor. The number of unknowns and equations increase with N. Also due to the high degree of symmetry in the problem for large N, knowledge of very high order terms in the series is required in order to determine the nature of the bifurcation. Therefore, a systematic use of normal form theory, the method of Lie transforms of Deprit(1968), and the special purpose algebraic processor **POLYPAK** written by the second author was necessary.

For N large there are more and more critical values of m which make the r.p.c.c. degenerate and all the cases that we investigated gave rise to a bifurcation and hence to new c.c. Figure 4 shows four of the eleven bifurcations which occur for the 13-body problem (12 around a regular 12-agon and 1 at the centroid).

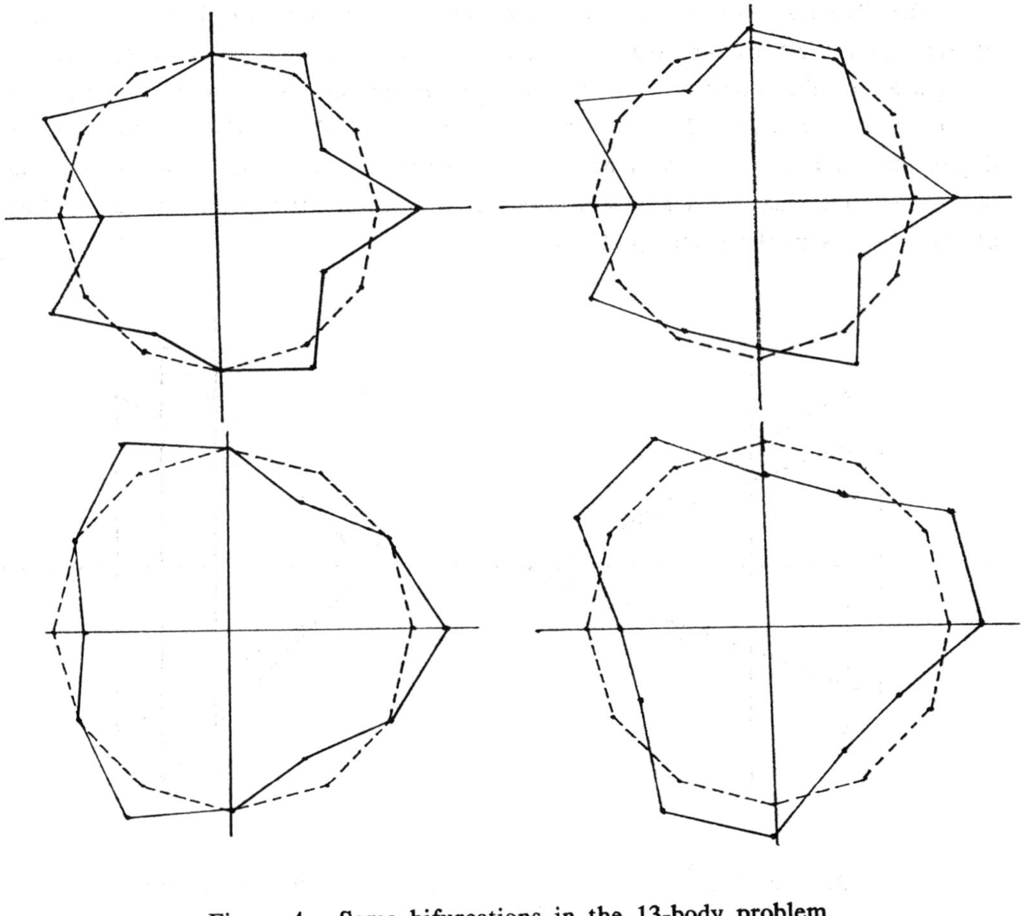

Figure 4. Some bifurcations in the 13-body problem.

References

Deprit, A. 1969: *Canonical transformation depending on a small parameter*, **Celestial Mechanics** 72, 173-79.

Dziobek, O. 1900: *Über einen merkwürdigen Fall des Vielkörperproblem*, **Astron. Nach.** 152, 32-46.

Euler, L. 1767: *De motu restilineo trium corporum se mutus attrahentium*, **Novi Comm. Acad. Sci. Imp. Petrop.** 11, 144-151.

Hoppe, R. 1879: *Erweiterung der bekannten Speziallösung des Dreikörper Problems*, **Archiv der Math. und Phy.** 64, 218-223.

Lagrange, J. L. 1772: *Essai sur le problem des trois corps*, **Oeuvers** 6, 272-292.

Lehmann-Filhes, R. 1891: *Über zwei Fälle des Vielkörpersproblem*, **Astron. Nach.** 127, 137-144.

Meyer, K. R. 1987: *Bifurcation of a central configuration*, **Celest. Mech.** 40, 273-282.

Meyer, K. R. and Schmidt, D. S.

1988a: *Bifurcations of relative equilibra in the 4 and 5 body problems*, **Ergod. Th. and Dynam. Sys.** 8*. 215-225.

1988b: *Bifurcations of relative equilibria in the N-body and Kirchhoff problems*, **SIAM J. Math. Anal.** 19(6), 1295-1313.

Moeckel, R. 1985: *Relative equilibria of the four-body problem*, **Ergod. Th. and Dyn. Sys.** 5, 417-435.

Moulton, F. R. 1910: *The straight line solutions of the problem of n-bodies*, **Bull. Amer. Math.** 13, 324-35.

Palmore, J. I.

1973: *Relative equilibria of the n-body problem*, Thesis, Univ. Calif., Berkeley.

1975: *Classifying relative equilibria II*, **Bull. Amer. Math. Soc.** 81(2), 489-91.

Saari, D. 1980: *On the role and properties of n-body central configurations*, **Celest. Mech. 21**, 9-20.

Schmidt, D. S. 1988: *Central configurations in* $\mathbb{R}^2$ *and* $\mathbb{R}^3$, **Hamiltonian Dynamical Systems**, Amer. Math. Soc., Providence, R.I., 59-76.

Simo, C. 1977: *Relative equilibrium solutions in the four body problem*, **Celest. Mech.** 18, 165-84.-84.

Kenneth R. Meyer
Department of Mathematical Sciences
University of Cincinnati
Cincinnati, Ohio 45221

Dieter S. Schmidt
Department of Computer Science
University of Cincinnati
Cincinnati, Ohio 45221

Contemporary Mathematics
Volume **108**, 1990

Leapfrogging of Vortex Filaments in an Ideal Fluid

M. S. Berger* and J. Nee*

The leapfrogging of vortex rings in an ideal fluid in three dimensions was first described by Helmholtz in his famous paper [1] on vortex motion. To describe this interaction of two thin vortex rings of equal strength propagating along a common axis, Love, in [2], found a quantitative analysis by assuming the two thin vortex rings could be represented by two pairs of two point vortices of equal strength placed symmetrically along a common axis. Leapfrogging in this case means that the relative motion of the two pairs of point vortices is periodic. Recently, Aref in [3] showed that the motion of 4 point vortices could be chaotic if the point vortices are not placed symmetrically. Here, in this paper, we demonstrate, for the first time, the leapfrogging phenomenon for two pairs of point vortices of different strengths but symmetrically placed.

We analyze this problem as a bifurcation process from an equilibrium for a Hamiltonian system. The usual approach of analyzing the geometry of level surfaces of the Hamiltonian seems difficult in this case.

1980 *Mathematics Subject Classification* (1985 *Revision*). 58A40, 58B15, 70H30, 70K40.
The final (detailed) version of this paper will be submitted for publication elsewhere.
*Research was partially supported by grants from the AFOSR and the NSF.

Section 1. The Physical Problem

Consider two pairs of two point vortices propagating along the x-axis (as common axis) initially symmetrically placed in $\mathbb{R}^2$ with Cartesian coordinates $P_1 = (x_1,y_1)$, $P_1' = (x_1,-y_1)$, $P_2 = (x_2,y_2)$ and $P_2' = (x_2,-y_2)$. These point vortices in $\mathbb{R}^2$ are to represent two pairs of cylindrical vortices of infinitely thin cross-section of different strengths. Here the circulation about the two point vortices of each pair are chosen equal and of opposite sign. The line of symmetry coincide for each pair (see Figure 1). Thus we choose the strength of $P_1 = \omega_1$, $P_1' = -\omega_1$, $P_2 = \omega_2$ and $P_2' = -\omega_2$. We set the ratio of vortex strengths $\frac{\omega_1}{\omega_2} = \lambda$. The case $\lambda = 1$ is classical as mentioned above. We show that if λ is any number between 1 and λ_c (a critical number approximately equal to .61623), the same leapfrogging phenomenon occurs as in the classic case.

In general for n point vortices with coordinates (x_i,y_i) $(i = 1,2,\ldots n)$ it is known that the motion of these n point vortices is governed by a Hamiltonian system

$$(1) \qquad \dot{z} = J\nabla H(z)$$

where $\underline{z} = (\underline{x}, y)$ and J is the matrix $\begin{bmatrix} 0 & -1 \\ 1 & 0 \end{bmatrix}$ and

$$H(x_i,y_i) = \sum_{i\neq j} \omega_i\omega_j \log\left[(x_i-x_j)^2 + (y_i-y_j)^2\right].$$

Thus, in the special case of the 4 symmetrically placed point vortices with strength $\alpha_1, -\alpha_1, \alpha_2, -\alpha_2$ we first find the conservation law

$$\alpha_1 y_1 + \alpha_2 y_2 = c \qquad c = \text{constant}$$

Consequently, the Hamiltonian H simplifies to (in this case)

$$H = \alpha_1\alpha_2 \log \frac{(x_1-x_2)^2 + (y_1+y_2)^2}{(x_1-x_2)^2 + (y_1-y_2)^2}$$

$$+ 2\alpha_1^2 \log 2y_1 + 2\alpha_2^2 \log 2y_2$$

We have supposed above that $\alpha_2 = \lambda\alpha_1$. Indeed, without loss of generality we suppose $\alpha_2 = 1$. Thus, to study relative motion we set

$$x = x_1-x_2 \qquad y = y_1-y_2$$

Thus the relative motion is described by the Hamiltonian system

(2)
$$\dot{x} = H_y(x, y, \lambda)$$
$$\dot{y} = - H_x(x, y, \lambda)$$

with $H = (1+\lambda)\left\{\log\left[x^2+(f+gy)^2\right]\left[x^2+y^2\right]^{-1} + \frac{1}{\lambda}\log(c+\lambda y) + \lambda \log(c-y)\right\}$

$$f = 2c(\lambda+1)^{-1} \quad \text{and} \quad g = (\lambda-1)(\lambda+1)^{-1}$$

where the Hamiltonian H above can be simplified to depend only on x, y and λ and the constant c defined above. In the sequel we shall demonstrate the leapfrogging of the two vortex pairs P_1P_1' and P_2P_2' on Figure 1 by proving

Theorem The system (2) has a smooth non-constant periodic solution for each fixed $\lambda\in(\lambda_0, 1)$, i.e. the relative motion of the vortex pairs is periodic. Here λ_0 is a critical parameter described in Section 2, and is approximately equal to .61623.

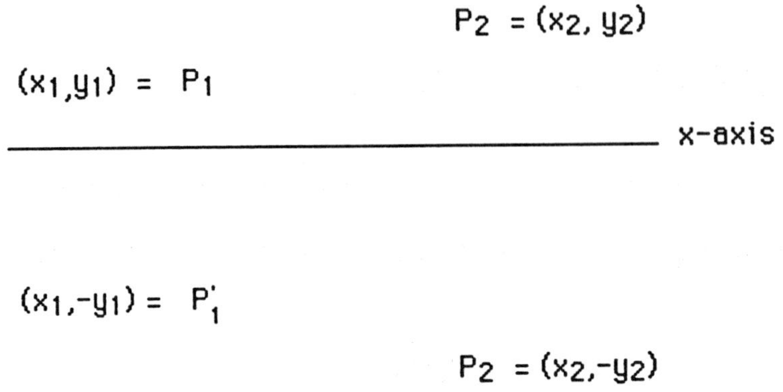

Figure 1 Coordinates for the 4 point vortices

Section 2 Analysis of the Hamiltonian System

We analyze the Hamiltonian system (2) by first showing that for λ in a large open interval $(\lambda_0, 1)$ that the system (2) has a unique equilibrium point $(0, y_\lambda)$ where y_λ is a real number implicitly defined by the unique real root of a cubic equation with two other complex conjugate roots. This analysis is carried out in an Appendix.

Then one writes the Hamiltonian system (2) in the form

$$\dot{z} = L(\lambda)z + O(|z|^2, \lambda) \tag{3}$$

for $L(\lambda) = H_{z\bar{z}}(0, y_\lambda)$ the Hessian of the Hamiltonian, near the singular point. One proceeds to show that the desired periodic solution of (3) bifurcates from this stationary point for each λ in the open interval $(\lambda_0, 1)$. To achieve this result we proceed by introducing a new parameter ω into (3) by changing the time scale $t = \omega s$. This allows the study of solutions of period ω in t

to be reduced to the study of 1-periodic solutions in s. The equation (3) becomes the nonlinear eigenvalue problem

$$\frac{dz}{ds} = \omega\left\{L(\lambda)z + O\left(|z|^2, \lambda\right)\right\} \tag{4}$$

To analyze the 1-periodic solutions of (4) it will suffice to consider the linearized problem about the stationary point $(0, y_\lambda)$

$$\frac{dZ}{ds} = \omega L(\lambda) Z \tag{4'}$$

<u>Section 3</u> <u>Analysis of the linearized problem (4') and the nonlinear problem (4)</u>

We consider now the periodic solutions $Z(s)$ of (4') in terms of (x, y) coordinates. Note first that $H_{xy}(0, y_\lambda) = 0$. Thus (4') simplifies to

$$\frac{d\tilde{x}}{ds} = \omega a(\lambda)\tilde{y} \tag{5}$$

$$\frac{d\tilde{y}}{ds} = -\omega b(\lambda)\tilde{x}$$

where a and b are constants depending on the parameter λ. Clearly we have periodic solutions for positive ω provided the product ab is positive. Indeed both $\tilde{x}(s)$ and $\tilde{y}(s)$ satisfy the second order equation

$$v_{ss} + \omega^2 a(\lambda) b(\lambda) v = 0$$

So $\tilde{x}(s)$ and $\tilde{y}(s)$ will be periodic solutions provided the product

$a(\lambda)b(\lambda)$ is positive. Using numerical analysis we find that the product $a(\lambda)b(\lambda) > 0$ provided λ is in the open interval $(.61626, 1)$. However, for $\lambda = \frac{1}{2}$ $a(\lambda)b(\lambda) < 0$, so if λ is outside this open interval, the result fails.

The nonlinear result will follow immediately from the standard Liapunov-Schmidt method for bifurcation theory, cf. Berger [4], once one restricts attention to periodic solutions of (4) and (4') that are odd in $x(s)$ and even in $y(s)$. Here we use the symmetry of H in (2), $H(x, y) = H(-x, y)$. This yields the fact 1-periodic solutions of (4') with the stated symmetry is one dimensional, a required prerequisite for the applicability of the standard theory. In order to utilize the Liapunov-Schmidt technique we simply note that there is a mapping A defined by (4) between the Hilbert spaces of periodic functions X and Y with

$$X = \left\{ \begin{array}{r} (x, y) \text{ of 1-periodic functions in } W_{1,2}\left[(0,1),\mathbb{R}^2\right] \text{ with } x(t) \text{ odd} \\ y(t) \text{ even} \end{array} \right\}$$

$$Y = \left\{ \begin{array}{r} (x, y) \text{ of 1-periodic functions in } L_2\left[(0,1),\mathbb{R}^2\right] \text{ with } x(t) \text{ even} \\ y(t) \text{ odd} \end{array} \right\}$$

To this mapping A the Liapunov-Schmidt method is applicable. Indeed $A(z) \equiv \frac{dz}{ds} - \omega\left[L(\lambda)z + 0\left(|z|^2, \lambda\right)\right]$ is a C' Fredholm mapping of index zero between the Banach spaces X and Y that can be used for the standard Liapunov-Schmidt theory.

The theory results in a nonzero-periodic solution $z_\omega(s)$ bifurcating from the equilibrium state $(0, y_\lambda)$ of (4) for each λ in the open interval $(\lambda_0,1)$. In fact these periodic solutions correspond to the nonconstant ω-periodic solutions of (2) as stated in the Theorem of Section 1.

Appendix

We write the Hamiltonian system (2) of Section 1 in the form

$$\dot{x} = h_1(x, y, \lambda)$$

$$\dot{y} = h_2(x, y, \lambda)$$

We prove

Lemma The equilibrium point of the system (2) is uniquely determined by the parameter λ in the open interval $(\lambda_0, 1)$.

Proof To determine the equilibrium points of (2) we note that

$$h_2(x, y, \lambda) = x\, c_\lambda(y)$$

where the smooth function $c_\lambda(y)$ vanishes at points where $h_1 \neq 0$. Thus the equilibrium points of (2) must be of the form $(0, y_\lambda)$ if they exist at all. We then determine the desired equilibrium points by solving

$$h_1(0, y, \lambda) = 0$$

Here y satisfies the equation

$$y^3 + \frac{3(\lambda-1)c}{\lambda^2+1}\, y^2 - \frac{6c^2}{\lambda^2+1}\, y + \frac{4c^3}{(\lambda^2+1)(\lambda-1)} = 0$$

Set $y_0 = cA$, then after simplifying the number A satisfies a cubic equation with coefficients depending only on λ. This equation has a unique real root A for each $\lambda \in (\lambda_0, 1)$.

Bibliography

1. H.Helmholtz, "On the integrals of the hydrodynamical equations that express vortex motion." Phil. Mag. (4) 33, pp. 485 - 512, 1887. English translation of 1858 German original.

2. E. H. Love, "On the motion of paired vortices with a common axis." Proc. London Math. Soc. , vol. 25, pp. 185 - 195, 1894.

3. H. Aref and S. Pomphrey, "Integrable and chaotic motions of four vortices." Proc. Royal Soc. London A, pp. 359 - 387, 1982.

4. M. S. Berger, Nonlinearity and Functional Analysis, Academic Press, New York, N.Y., 1977.

Center for Applied Mathematics
University of Massachusetts, Amherst

Contemporary Mathematics
Volume **108**, 1990

CALCULATION OF SHARP SHOCKS USING SOBOLEV GRADIENTS

J. W. Neuberger[1]

Introduction. This note concerns the use of a descent method to calculate sharp shocks in flow problems. This is a pressing problem in computational fluid dynamics. Many viscosity-free flow problems have a multitude of solutions only a few of which are physically admissible. In function space the addition of a viscosity term commonly leads to problems with a unique solution, but a severe difficulty typically arises in numerical treatments. Very roughly, this difficulty arises in the following way: For computationally realistic grids, a physically excessive amount of viscosity is required in order to achieve a solution. A rather badly smeared shock is typically then found – a shock smeared too much for design purposes. If viscosity is small enough to be physically reasonable, then numerically a solution process typically behaves much as though there were no viscosity term present. With such small viscosity, one is likely to calculate physically unreasonable solutions. A general mathematical reference for flow problems is [8]. For additional background concerning computational practice on flow problems see the early paper [4] and [2].

In the present note we try to offer a way around this difficulty. Sobolev gradients described in [6],[7] provide a point of departure. Some evidence is given for the numerical viability of these gradients and consequently the new one presented in this note. Finally it is shown that our methods produce a physically correct solution to a one dimensional flow problem. See [1] for background on steepest descent.

We now present some background about gradients. For n a positive integer and ϕ a real-valued $C^{(1)}$ function on R^n, it is customary to take for a gradient of ϕ:

$$(1)\quad (\nabla\phi)(x_1,..,x_n) = (\phi_1(x_1,..,x_n),..,\phi_n(x_1,..,x_n)),\ (x_1,..,x_n) \in R^n,$$

1980 *Mathematics Subject Classification* (1985 *Revision*). 65M10, 65K10, 49D10, 76H05, 35A40.
The final (detailed) version of this paper will be submitted for publication elsewhere.
[1] Supported by Texas Advanced Research Program.

where ϕ_i denotes the partial derivative of ϕ in its ith argument. We note that $(\nabla\phi)(x)$ is the member h of R^n so that $\phi'(x)h$ is maximum subject to the condition

$$\|h\|_{R^n} = |\phi'(x)|$$

where $\phi'(x)$ denotes the Frechet derivative of ϕ at $x \in R^n$ and $|\phi'(x)|$ denotes the norm of $\phi'(x)$ regarded as a member of the dual of R^n. Note also that one has the identity

$$\phi'(x)h = \langle h,(\nabla\phi)(x)\rangle_{R^n}$$

where $\| \|_{R^n}$ and $\langle\ ,\ \rangle_{R^n}$ denotes the usual norm and inner product respectively on R^n.

Now for any Hilbert space H and any member $f \in H^*$ (the dual space of H) there is a unique member $y \in H$ such that

$$f(x) = \langle x,y\rangle_H,\ x \in H.$$

In particular, if one has two inner products on the same finite dimensional linear space, then for a given linear functional on this space one has distinct representations relative to these two inner products. This observation has substantial numerical consequences as is illustrated by what follows.

Denote by n a positive integer and denote by H the space of all (n+1)-tuples of the form $y = (y_0,y_1,\ldots,y_n)$ so that $y_0 = 0$. On the space H define two norms:

$$\|y\|_1 = (\Sigma_{i=1}^n y_i^2)^{1/2},\ \|y\|_2 = [\Sigma_{i=1}^n((y_i-y_{i-1})^2/\delta + (y_i+y_{i-1})^2/2)]^{1/2},$$

$y = (y_0,y_1,\ldots,y_n) \in H$, $\delta \equiv 1/n$. Try to make the problem of approximating a solution f to

$$(2) \qquad f' = f,\ f(0) = 1 \quad \text{on} \quad [0,1]$$

into an optimization problem using the function ϕ:

$$\phi(y) = \Sigma_{i=1}^n[((y_i+1)-(y_{i-1}+1))/\delta + ((y_i+1)+(i_{i-1}+1)/2]^2/2.$$

Clearly if $y \in H$ and $\phi(y) = 0$, then the vector $y + w$, $w \equiv (1,\ldots,1)$, provides a second order approximation to a solution to (2).

The Frechet derivative $\phi'(y)$, $y \in H$, has a representation in each of the inner products equivalent to $\| \|_1$ and $\| \|_2$, respectively:

$$\phi'(y)h = \langle h,(\nabla_1\phi)(y)\rangle_1;\ \phi'(y)h = \langle h,(\nabla_2\phi)(y)\rangle_2.$$

The second of these expressions serves as a defining relationship for

$(\nabla_2\phi)(y)$. The first may be used as a defining relationship for $(\nabla_1\phi)(y)$ or quivalently this gradient may be defined using (1).

Theorem. If $y \in H$ and $\phi(y) \neq 0$, then

$$(\inf_{\alpha>0}\phi(y - \alpha(\nabla_2\phi)(y)))/\phi(y) \leq .4 \quad .$$

A proof is found in [5].

By contrast, using $\| \|_1$ to generate a gradient for ϕ, no such estimate is possible. From [7],[5], it follows that for each positive integer n there is a smallest possible $c_n < 1$ (n being the number of pieces into which $[0,1]$ is broken) so that for all $y \in H$ with $\phi(y) \neq 0$,

$$(\inf_{\alpha>0}\phi(y-\alpha(\nabla\phi)(y)))/\phi(y) \leq c_n.$$

However, $\lim_{n\to\infty} c_n = 1$. The constant 1 in such inequalities is indicative of no progress at all in decreasing ϕ.

Actual computation confirms that steepest descent with $\nabla_2\phi$ works very well (with about seven iterations usually sufficing) whereas for $\nabla_1\phi$ the number of iterations required increases drastically as n increases.

We now illustrate how a gradient constructed with respect to $\| \|_2$ can be calculated in terms of a gradient constructed with respect to $\| \|_1$. We emphasize that $\nabla_2\phi$ is not calculated by means of taking partial derivatives as in (1). This will provide some motivation for later developments. The gradient $\nabla_1\phi$ may be calculated by simply taking the necessary partial derivatives of ϕ since $\| \|_1$ is essentially the usual R^n norm. To calculate $\nabla_2\phi$, we introduce D by means of $Dy = \binom{Ky}{D_1y}$ where for $y = (y_0,y_1,\ldots,y_n) \in H$,

$$Ky = ((y_1+y_0)/2,\ldots,(y_n+y_{n-1})/2),$$

$$D_1y = ((y_1-y_0)/\delta,\ldots,(y_n-y_{n-1})/\delta), \quad \delta \equiv 1/n.$$

Then for $h,y \in H$, $\phi'(y)h = \langle h,(\nabla_1\phi)(y)\rangle_1 = \langle h,(\nabla_2\phi)(y)\rangle_2 =$

$$\langle Dh,D(\nabla_2\phi)(y)\rangle_{R^{2n}} = \langle h,D^*D(\nabla_2\phi)(y)\rangle_1 = \langle h,\pi_0D^*D(\nabla_2\phi)(y)\rangle_1$$

where $\pi_0(x_0,x_1,\ldots,x_n) = (0,x_1,\ldots,x_n)$, $(x_0,x_1,\ldots,x_n) \in R^{n+1}$, i.e., where π_0 is the orthogonal projection of R^{n+1} onto H.

Hence it must be that if $E \equiv \pi_0 D^* D|_{R(\pi_0)}$, then

$$E(\nabla_2\phi)(y) = (\nabla_1\phi)(y),\ y \in H.$$

Therefore

$$(\nabla_2\phi)(y) = E^{-1}(\nabla_1\phi)(y).$$

One may observe that E corresponds to a positive definite symmetric matrix and its inverse surely exists.

Much of our development so far is presented in more general form in [5],[6],[7] and has been presented here as background and motivation for the material in the remainder of this note.

For some (but not all) Banach spaces H, if $f \in H^*$, the dual of H, then there is a unique element $h \in H$ so that

$$fh \text{ is maximum subject to } |f|_{H^*} = \|h\|_H.$$

One such space of interest here is H where for fixed positive integers n,p, $p \geq 2$, H is the space of all $y = (y_0, y_1, \ldots, y_n)$ so that

$y_0 = 0 = y_n$ and

$$\|y\|_H = [\Sigma_{i=1}^n(|(y_i - y_{i-1})/\delta|^p + |(y_i + y_{i-1})/2|^p)]^{1/p},\ \delta \equiv 1/n.$$

Suppose $f \in H^*$ and denote by $z \in H$ an element so that

$$fy = \langle y,z\rangle_{R^{n+1}} \text{ for all } h \in H.$$

We address the problem of maximizing fh subject to the constraint that $\|h\|_H$ be a given constant c.

Define $\beta(y) = \|y\|_H^p/p$, $y \in H$. Define $Q(x) = p|x|^{p-2}x$, $x \in R$ and note that since

$$\beta(y) = [\|Ky\|^p_{R_p^{n+1}} + \|D_1 y\|^p_{R_p^{n+1}}],\ y \in H,$$

it follows that

$$\beta'(y)h = \langle Q(Dy), Dh\rangle_{R^{n+1}} = \langle \pi D^* Q(Dy), h\rangle_{R^{n+1}},$$

and so the conventional gradient $\nabla\beta$ of β is given by

$$(\nabla\beta)(y) = E_p(y),\ y \in H$$

where $\pi(y_0, y_1, \ldots, y_n) = (0, y_1, \ldots, y_{n-1}, 0)$, $y \in R^{n+1}$, and

$$E_p(y) = \pi D^*(Q(Dy)),\ y \in H.$$

By the theory of Lagrange multipliers, for $y \in H$ to maximize fh subject to the given constraint $\beta(y) = c$, it must be that

$(\nabla\beta)(y)$ and z are linearly dependent. In other words, y must be a positive constant times $E_p^{-1}z$ where z represents f as indicated above, the constant to be used being chosen so that the gradient has the correct norm.

This development generalizes to numerical versions of various Sobolev spaces $H^{m,p}(\Omega)$, $\Omega \subset R^k$. k,m,p positive integers with $p \geq 2$. For $m = 1$, $n \geq 2$, the analogue of E above is a numerical version of the p-Laplacian. There is extensive information on this setting for the case $p = 2$ in [6],[7]. It is expected that a systematic generalization of some of these results for the case $p > 2$ will be published elsewhere.

In the remaining part of this note we illustrate how the above may be applied to the problem of calculating sharp shocks in a flow problem. Our illustration centers about a nozzle problem from [3].

Suppose that an nozzle, symmetric about its axis, of length 2 and has a cross-sectional area

$$A(x) = .4[1 + (1-x^2)], \quad 0 \leq x \leq 2.$$

In [3] it is supposed that pressure and velocity depends only on the distance along the axis of the nozzle and that pressure is related to velocity by

$$p(u) = [1 + ((\gamma-1)/2)(1-u^2)]^{\gamma/(\gamma-1)}$$

for all velocities u so that $1 + ((\gamma-1)/2)(1-u^2) \geq 0$ where $\gamma = 1.4$ (for air). Define a density function m by $m(u) = -p'(u)$ for all relevant u. Further define

$$J(f) = \int_0^2 Ap(f'), \quad f \in H^{1,7}.$$

We choose $H^{1,7}$ since $f \in H^{1,7}$ implies $p(f') \in L_1([0.2])$ with $\gamma = 1.4$. Take a first variation:

$$(3) \qquad J'(f)h = -\int_0^2 Am(f')h', \quad f,h, \in H^{1,7}$$

where the perturbation h is required to satisfy $h(0) = 0 = h(2)$ and f is required to satisfy $f(0) = 0$, $f(2) = c$, a given fixed number.

For a numerical analogue of the above, break $[0,2]$ into k pieces of equal length. Denote by H the finite dimensional analogue of $H^{1,7}([0,2])$ indicated above. Denote by J_k the numerical analogue of J for this partition.

Denote by H_0 the set of all $u \in H$ so that $u = (u_0,u_1,\ldots,u_k)$, $u_0 =$

$0 = u_k$. Determine $h \in H_0$ so that $J_k'(u)h$ is maximum subject to $\|h\|_H = |J_k'(u)|_H^*$. Define $F(u)$ to be equal to this last choice of maximizing h.

Finally define

$$\phi(u) = \|F(u)\|_H^7/7, \quad u \in H_0.$$

Then for $u \in H_0$, calculate $h \in H_0$ so that $\phi'(u)h$ is maximum subject to the requirement that $\|h\|_H = |\phi'(u)|_H^*$ and denote this maximizing h by $(\nabla\phi)(u)$.

Consider now the iteration

$$(4) \qquad u \to u - \alpha(\nabla\phi)(u)$$

where α is chosen optimally in each step.

Before presenting results of our calculations we point out some features of the problem which themselves are at the heart of many difficulties in computational fluid dynamics.

Returning to equation (3), we see, after integrating by parts, that $J'(f)h = 0$ for all admissible $h \in H$ if and only if

$$(5) \qquad (Am(f'))' = 0.$$

This seemingly simple ordinary differential equation is essentially a first order equation since f does not appear explicitly. Assuming for the moment that a solution f to (5) has a second derivative, we calculate

$$(6) \qquad A'm(f') + Am'(f)f'' = 0$$

and so we see that the equation is possibly singular since it may be that $m(f')$ has a zero somewhere in $[0,2]$. An easy calculation yields that (6) is singular at a number t provided that $f'(t) = 1$. Our interpretation of f' is as a velocity. According to [3], velocity one corresponds to the speed of sound.

On the other hand an examination of m yields that for each choice of a value for $y \in [0,1)$, there are precisely two values $x_1, x_2 \in [0,((\gamma+1)/(\gamma-1))^{1/2})$ so that $x_1 < x_2$ and $m(x_1) = y = m(x_2)$. The value x_1 corresponds to a subsonic velocity and the value x_2 corresponds to a supersonic velocity. So for a solution f of (4) for which (i) $f'(t_1) < 1$ for some t_1 and $f'(t_2) > 1$ for some t_2, and (ii) the boundary condition $f(0) = 0$, $f(2) = c$ holds, we may construct many other solutions as follows:

Pick two subintervals $[a,b]$ and $[c,d]$ of $[0,2]$ so that f is subsonic on $[a,b]$ and supersonic on $[c,d]$. Adjust f' up to supersonic values on a part of [a,b] and adjust f' down to subsonic values on a part of [c,d] in such a way that $\int_0^2 f'$ does not change.

In this way one constructs from a single solution f a wide family of solutions. However, of all the solutions (under the given conditions) to (5) (i.e., critical points of the functional J), there is just one which is physically admissible: The one for which there is no jump in f' from subsonic to supersonic in the streamline direction.

We raise the question as to how a descent method such as we are proposing can possibly pick out this one physically correct solution from among the large collection of mathematical ones.

An answer we propose is the following: Take the Euler–Lagrange equation for the variational problem. Form a new equation

$$-(Am(f'))' + \epsilon f''' = 0$$

under appropriate boundary conditions. This problem no longer has the possibility of being singular. Use the solution to this viscosity equation to start the descent process as indicated above. The smooth graph in the figure is a viscosity solution (viscosity = .1). The graph with the sharper drop is the resulting viscosity–free solution obtained by our method.

From [6] and [7] one has that under appropriate situations in which there are many solutions, that steepest descent tends to converge to the nearest (in perhaps some non–euclidean sense) to the starting estimate. On the basis of evidence such as is presented here it is natural to conjecture that the "nearest" viscosity–free solution to a viscosity solution (for a reasonable small viscosity) is a good approximation to the physically correct solution. Numerical experiments point to the truth of this conjecture in higher dimensional problems, but much theoretical and experimental work remains to be done.

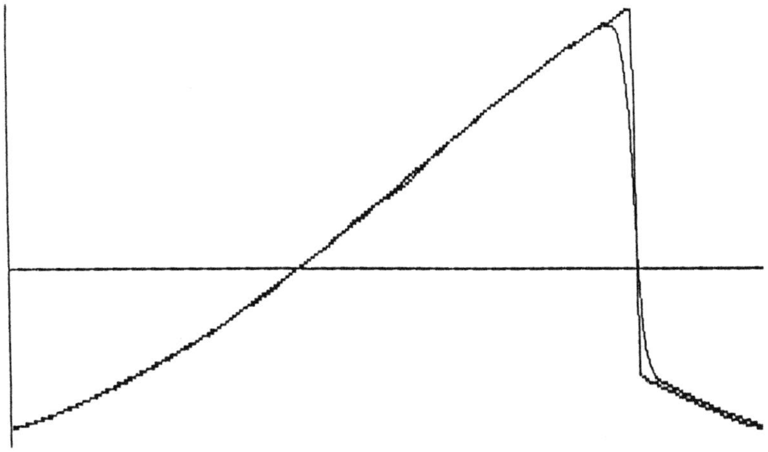

References

1. M. Berger, Nonlinearity and Functional Analysis, Academic Press (1977).

2. A. Jameson, Transonic Flow Calculations, Numerical Methods in Fluid Dynamics, H. Virz and J. J. Smoller, eds, McGraw Hill (1978)

3. F. J. Johnson, Application of a Poisson Solver to the Transonic Full Potential Equations, Boeing Note #AERO-B8113-C81-004.

4. E. M. Murman and J. D. Cole, Calculation of Plane Steady Transoni Flows, AIAA J. (9) (1971), 114-121.

5. J. W. Neuberger, Steepest Descent for General Systems of Linear Differential Equations in Hilbert Space, Springer Lecture Notes 1032 (1983), 390-406.

6. ...., Steepest Descent and Differential Equations, J. Math. Soc. Japan 37 (1985), 187-195.

7. ...., Constructive Variational Methods for Differential Equations, J. Nonlinear Analysis 13 (1988), 413-428.

8. J. J. Smoller, Shock Waves and Reaction-Diffusion Equations, Springer-Verlag (1983).

Department of Mathematics
University of North Texas
Denton, TX 76203

and

Formal Systems Design and Development, Inc.
158 N. Ross
Auburn, AL 36830

Contemporary Mathematics
Volume **108**, 1990

MONODROMY PRESERVING DEFORMATION OF THE DIRAC OPERATOR ACTING ON THE HYPERBOLIC PLANE [1, 2]

John Palmer[3] and Craig A. Tracy[4]

ABSTRACT. We extend the monodromy preserving deformation theory of Sato, Miwa, and Jimbo of the Euclidean Dirac operator acting on $\mathbf{R}^2$ to the Dirac operator acting on the Poincaré upper half-plane $\mathbf{H}$. The hyperbolic Laplace transform of our extended equations is a Fuchsian system.

§**1. Introduction.** The notion of monodromy for linear differential equations goes back to Riemann's study of the hypergeometric equation. Linear differential equations on the Riemann sphere with meromorphic coefficients have solutions with branch type singularites at the singular points. Local fundamental solutions to such equations are determined up to a constant matrix multiplier and one can study the branching behavior of the solutions by examining the *monodromy matrices* which arise in the analytic continuation of such fundamental solutions about the singular points. Suitably defined, the monodromy matrices generate a representation of the fundamental group of the sphere punctured at the singular points. Riemann's investigation of the hypergeometric equation showed the power of thinking about solutions to linear equations in these terms and he further proposed the problem of trying to characterize linear differential equations with regular singular points in terms of their monodromy group. A version of this problem found its way onto Hilbert's famous list at the turn of the century and it has since been known as the Riemann-Hilbert problem. Plemelj, Hilbert, and Birkhoff offered solutions to this problem in the early years of this century and in 1957 Röhrl [21] published an account which generalized to Riemann surfaces and exhibited a connection between this problem and the problem of trivializing holomorphic vector bundles on Riemann surfaces. One can find a comprehensive review of modern developments on Hilbert's 21st problem in the article by Katz [9].

We drop the Riemann-Hilbert problem at this point to pick up the threads of a related development: Monodromy preserving deformations of linear differential equations.

---

[1] *Mathematics Subject Classification* (1985 *Revision*). Primary 35Q15, 81E; Secondary 81E40, 82A69.
[2] The final version of this paper will be submitted for publication elsewhere.
[3] Supported in part by the National Science Foundation, Grant No. DMS–87–03169.
[4] Supported in part by the National Science Foundation, Grant No. DMS–87–00867.

In 1907 R. Fuchs [3] considered the problem of varying the singular points of a second order equation with five regular singular points so as to preserve the monodromy group of the resulting equation. He discovered that when three of the singular points were fixed the deformation problem determined one of the two remaining singular points as a Painlevé function of the other. The Painlevé transcendents are solutions to non-linear differential equations in the complex plane which are characterized by a property which generalizes the behavior of Riccati equations. The only "movable" singularities of the solutions to such equations are poles (roughly, "movable" singularities are ones that depend on initial conditions but do not reflect manifest singularities in the coefficients of the equation itself). P. Painlevé classified the second order equations of this type in work which is summarized in Ince's book on differential equations [8] (cf. [5], [15]). In 1912 Garnier [4] generalized Fuch's work to N regular singular points and in the same year Schlesinger [23] generalized it to systems.

Schlesinger considered systems of the form

$$\frac{dY}{dz} = \sum_{j=1}^{p} \frac{A_j}{z - a_j} Y ,$$

in which the point at infinity is a regular point and the A_j are $n \times n$ complex matrices. He posed the problem of determining how the coefficients A_j must vary as functions of the a_j ($j = 1, \ldots, p$) so that the monodromy group of the resulting equation does not depend on the points $\{a_1, \ldots, a_p\}$. He found that

$$dA_j = -\sum_{k \neq j} [A_j, A_k] \frac{da_j - da_k}{a_j - a_k} , \tag{1.1}$$

an equation which is now known as the *Schlesinger equation*. In the same paper Schlesinger claimed to prove that the solutions to these equations have the Painlevé property (the only movable singularities are poles). In the early 1920's various special cases of the Schlesinger equations were integrated in terms of the Painlevé transcendents. A recent account of such results can be found in Okamoto [15].

We would next like to trace some developments that emerged (in a somewhat indirect fashion) from work of Latta [10] and Myers [14]. They found that a certain integral equation which arises in scattering from a strip has an intimate connection with a Painlevé equation of the third kind. In their study of the scaling limit of the two-point correlation function of the two-dimensional Ising model, Wu, McCoy, Tracy, and Barouch (referred to as WMTB henceforth, see also [12]) encountered this same integral equation. They were able to make use of the Latta–Myers work to express the scaled correlation in terms of a Painlevé transcendent [27]. To someone aware of the intimate relation between Painlevé transcendents and monodromy preserving deformation theory, the WMTB result suggests that "monodromy" is lurking somewhere in the continuum limit of the Ising model. In the years 1977–1980 Sato, Miwa, and Jimbo [22] developed their theory of *Holonomic Quan-*

tum Fields and unveiled the "monodromy connection" in the the WMTB result for the Ising model which is understood as a special case of their more general theory. Roughly speaking the holonomy fields which are the central objects in their theory can be understood to "create monodromy" in the solutions to a linear differential equation. There are many novel features in this work, but the one which we wish to mention first is that in generalizing the WMTB Ising model results, Sato, Miwa, and Jimbo (referred to as SMJ henceforth) were lead to consider monodromy preserving deformations of the Euclidean Dirac operator (the elliptic version of the hyperbolic operator originally introduced by Dirac) on $\mathbf{R}^2$. This operator is the analogue of the Cauchy-Riemann operator, $\bar{\partial}_z$, in the developments described above for the Riemann-Hilbert problem; it is not the analogue of the linear differential equation with regular singular points! The object in their theory which generalizes the two-point function of the Ising model is the vacuum expectation of a product of holonomy fields, which they refer to as a *τ-function*. Each holonomy field $\varphi(a,l)$ depends on a point $a \in \mathbf{R}^2$ and parameter ℓ which determines the monodromy parameter $\lambda = e^{2\pi i \ell}$. Let $\{a_1, \ldots, a_p\}$ denote a collection of points in $\mathbf{R}^2$ and let $\ell_j \in \mathbf{C}$ for $j = 1, \ldots, p$. Write

$$\tau(a) := < \varphi(a_1, \ell_1) \cdots \varphi(a_p, \ell_p) >$$

for the vacuum expectation of a product of holonomy fields. The result of SMJ which generalizes the WMTB result is that $\tau(a)$ can be expressed in terms of a solution to a monodromy preserving deformation equation for the Dirac equation with monodromy $\lambda_j = e^{2\pi i \ell_j}$ at a_j. To obtain this result SMJ developed a large body of new results in the representation theory of Clifford algebras and the spin extension of the orthogonal group, invented the notion of holonomy fields and monodromy preserving deformations of the Dirac equation, and made novel use of the notion of local operator product expansions from quantum field theory.

We are interested in describing here the first steps in a program to generalize the SMJ analysis to a family of geometrically natural operators acting on the Poincaré upper half-plane $\mathbf{H} := \{x + iy : y > 0\}$ including the appropriate versions of the Helmholtz and Dirac operators. We view the problem of making such a generalization in two parts:

(1) Generalize the monodromy preserving deformation theory to the Poincaré upper half-plane.
(2) Identify the holonomy fields or otherwise characterize the τ-function for the Poincaré upper half-plane problem.

Problem (1) has been solved for the Helmholtz equation [25] and in the next two sections of this note we will describe analogous results for the Dirac equation. The crucial observation in pushing through the deformation analysis is that $\mathbf{H}$ can be thought of as a homogeneous space for the group $SL_2(\mathbf{R})$, i.e. $\mathbf{H} = SL_2(\mathbf{R})/SO(2)$. This group acts on the space of solutions to the Helmholtz (or Dirac) equation and the infinitesimal version of this fact has consequences for the monodromy problem which can be exploited in much the same fashion that SMJ exploit the Euclidean motion group in $\mathbf{R}^2$ (note that $\mathbf{R}^2 = E(2)/O(2)$). One surprise (at least at our current level of understanding) is the role that the hyperbolic Laplace transform plays in connecting deformation theory on $\mathbf{H}$ with a deformation problem in the plane for the Cauchy-Riemann operator.

We have made less progress in solving problem (2). In the remainder of this introduction we will try to explain the nature of this problem. One reason for difficulty in problem (2) is that even in the original SMJ analysis there is no mathematically satisfactory definition of a holonomy field. These quantum fields are very singular objects and SMJ do not have a definition for which one can prove the existence of the correlation functions at all points for which the field arguments do not coincide ($a_j \neq a_k$ for $j \neq k$ in $\tau(a)$ above). This problem has been overcome in the Ising model case by rigorously controlling the scaling limit of the correlations on the lattice [19] and a similar lattice approximation for more general holonomy fields [16] allows one to define rigorously holonomy field correlations and to recover all the rest of the SMJ analysis [1]. While this is instructive, the numerous extra complications on the lattice make it clear that one has sacrificed much formal simplicity to make sense of the holonomy correlations in this fashion. A lattice approximation to the upper half-plane problem does not seem the right way to go.

We are interested in holonomy fields mainly because one understands what the τ-function is in terms of these fields. If one could formulate an independent notion of what the τ-function is perhaps one could by-pass the problem of constructing the holonomy fields. We will now briefly describe one success along these lines. If one returns to the original Ising problem then it is interesting to observe that the correlation functions have long been understood to be Pfaffians (cf. [13]). In [17] it is shown that one may rigorously interpret such a Pfaffian as the Pfaffian of a linear difference operator on $\ell^2(\mathbf{Z}^2)$ with inhomogeneities along "branch cuts" that emerge from the locations of the spin sites. Intuitively, what happens in the continuum limit is that the inhomogeneities on the lattice force genuine "branch cuts" in the domain of the resulting operator. The τ-function might then be understood as the Pfaffian of the Dirac operator with a domain that reflects prescribed branching (-1 for the Ising model) at the points a_j. This picture has not yet been worked out in the Dirac case but it has been worked out for the Cauchy-Riemann operators associated with the Riemann-Hilbert problem [11], [18]. In [11] Malgrange shows how to give an independent geometric interpretation for the τ-function which SMJ associate to the Riemann-Hilbert problem (in addition to a beautiful proof of the Painlevé property for solutions to the Schlesinger equations). In [18] it is shown that one may reinterpret the Malgrange analysis to get

$$\tau(a) = det(\bar{\partial}_{a,L}) ,$$

where $\bar{\partial}_{a,L}$ is a "Cauchy-Riemann operator" whose domain incorporates prescribed branching ($exp(2\pi L_j)$) at the points a_j. The determinant bundle introduced by Quillen [20] and the det^* bundle introduced by Segal and Wilson [24] together with some ideas in Witten [26] allow one to make sense of this determinant in rigorous fashion. Furthermore we believe that taking $\bar{\partial}_{a,L}$ to be the fundamental object in the theory clarifies many of the constructions in the SMJ analysis (the Green's function for this operator does make an appearance in the SMJ work but it never quite makes it to center stage in the later work on the Dirac operator). In any case the generalization of the Malgrange analysis to the Dirac case looks quite promising and we believe this points the way toward the introduction of τ-functions in the Poincaré upper half-plane problem.

§2. Dirac Operator in Hyperbolic Plane and the Isomonodromy Problem. Recall $\mathbf{H} := \{z = x + iy : y > 0\}$ denotes the Poincaré upper half-plane, and that the measure $d\mu := y^{-2}\,dxdy$ is invariant under the isometries of $\mathbf{H}$ where an isometry $\gamma = \begin{pmatrix} a & b \\ c & d \end{pmatrix} \in SL_2(\mathbf{R})$ acts on $z \in \mathbf{H}$ by

$$\gamma z = \frac{az+b}{cz+d} .$$

On $\mathbf{H}$ a spin connection can be defined, and hence a Dirac operator (cf. [6]) acting on spinors $\Psi = \begin{pmatrix} \psi_1 \\ \psi_2 \end{pmatrix}$:

$$\Gamma_k \Psi := \begin{pmatrix} m & -K_k \\ -L_{k+1} & m \end{pmatrix} \begin{pmatrix} \psi_1 \\ \psi_2 \end{pmatrix} = 0 \tag{2.1}$$

where we have written the Dirac operator Γ_k in terms of the *Maass operators* (cf. [2])

$$K_k := (z - \bar{z})\partial + k ,$$
$$L_k := (\bar{z} - z)\bar{\partial} - k ,$$

$k \in \mathbf{R}$ ($k = -\frac{1}{2}$ corresponds to spin one-half), $m > 0$ is the mass, and $\partial := \frac{1}{2}(\partial_x - i\partial_y)$, $\bar{\partial} := \frac{1}{2}(\partial_x + i\partial_y)$. Introducing the Laplacian

$$D_k := y^2(\partial_x^2 + \partial_y^2) - 2iky\partial_x ,$$

it is straightforward to verify

$$\begin{aligned} D_k &= L_{k+1}K_k + k(k+1) , \\ D_{k+1} &= K_k L_{k+1} + k(k+1) , \end{aligned} \tag{2.2}$$

and

$$D_{k+1}K_k = K_k D_k ,$$
$$D_k L_{k+1} = L_{k+1} D_{k+1} .$$

If we introduce $s > 1$ by $s(s-1) := m^2 + k(k+1)$, then using (2.2) we see each component ψ_i (i=1,2) of the Dirac equation (2.1) satisfies

$$\begin{aligned} D_{k+1}\psi_1 &= s(s-1)\psi_1 , \\ D_k\psi_2 &= s(s-1)\psi_2 . \end{aligned} \tag{2.3}$$

Let $\{a_1, \ldots, a_n\}$ denote distinct points in $\mathbf{H}$ and fix $-\frac{1}{2} < \ell_\nu < \frac{1}{2}$ ($\nu = 1, \ldots, n$). Denote by A and L the $n \times n$ diagonal matrices with diagonal elements $A_{\nu\nu} = a_\nu$ and

$L_{\nu\nu} = \ell_\nu$, and let $R_\nu(\theta)$ denote the rotation by θ (in the counterclockwise direction) about a_ν. We now introduce the fundamental vector space of "multivalued wave-functions":

Definition 2.1. *$W_k(L, A)$ will denote the complex linear space of multivalued solutions Ψ to the Dirac equation (2.1) in $\mathbf{H} - \{a_1, \ldots, a_n\}$ satisfying:*

$$\Psi\big(R_\nu(2\pi)(z,\bar{z})\big) = exp\big(2\pi i(\ell_\nu - \frac{1}{2})\big)\Psi(z,\bar{z})$$

for z near a_ν , and

$$I(\Psi,\Psi) := \frac{m}{4}\int_{\mathbf{H}} d\mu\,(\psi_1\bar{\psi}_1 + \psi_2\bar{\psi}_2) < \infty\,.$$

To study the behavior of $\Psi \in W_k(L, A)$ near a_ν, we need to introduce new coordinates. For a given point $z_0 \in \mathbf{H}$ we define the *geodesic polar coordinates* $\big(r(z,z_0), \theta(z,z_0)\big)$ centered at z_0 by

$$\frac{z - z_0}{z - \bar{z}_0} = exp\big(i\theta(z,z_0)\big)\tanh\big(\frac{1}{2}r(z,z_0)\big)\,.$$

Often times we abbreviate the geodesic polar coordinates to (r, θ). Following Fay [2] it is useful to introduce a unitary map $\Lambda_{k,\nu} : W_k(L, A) \to W_{k,\nu}(L, A)$ defined by

$$\begin{aligned}\check{\Psi}_\nu(z,\bar{z}) &= \Lambda_{k,\nu}\Psi(z,\bar{z})\\ &= \begin{pmatrix}\Lambda_\nu^{k+1} & 0\\ 0 & \Lambda_\nu^k\end{pmatrix}\begin{pmatrix}\psi_1\\ \psi_2\end{pmatrix}\end{aligned}$$

where

$$\Lambda_\nu = \frac{z - \bar{a}_\nu}{a_\nu - \bar{z}}\,.$$

Separating variables in geodesic polar coordinates, we can show that if $\Psi \in W_k(L, A)$, then the *local expansions* of Ψ are of the form:

$$\check{\Psi}_\nu(z,\bar{z}) = \sum_{\ell=\ell_\nu,\ell_\nu+1,\ldots} c_\nu(\ell,+)\,\check{w}_{1\nu}(z,\bar{z};\ell;k) + \sum_{\ell=\ell_\nu,\ell_\nu-1,\ldots} c_\nu(\ell,-)\,\check{w}_{2\nu}(z,\bar{z};\ell;k)$$

for z near a_ν, and

$$\check{\Psi}_{z_0}(z,\bar{z}) = \sum_{\ell\equiv\ell_\infty[\mathbf{Z}]} c_\infty(\ell)\,\check{w}_\infty(z,\bar{z};\ell;k)\,,$$

for z near ∞, where $\ell_\infty - 1/2 = \ell_1 + \ldots + \ell_n - n/2$ (there will be no monodromy at ∞ if $\ell_\infty \equiv \frac{1}{2}[\mathbf{Z}]$) and

$$\begin{aligned}\check{w}_{1\nu}(z,\bar{z};\ell;k) &:= P^{-\ell+\frac{1}{2}}_{s,-k-1}(z,a_\nu)e_1(\theta,\ell) + mP^{-\ell-\frac{1}{2}}_{s,-k}(z,a_\nu)e_2(\theta,\ell)\,,\\ \check{w}_{2\nu}(z,\bar{z};\ell;k) &:= mP^{\ell-\frac{1}{2}}_{s,k+1}(z,a_\nu)e_1(\theta,\ell) + P^{\ell+\frac{1}{2}}_{s,k}(z,a_\nu)e_2(\theta,\ell)\,,\\ \check{w}_\infty(z,\bar{z};\ell;k) &:= mQ^{\ell-\frac{1}{2}}_{s,k+1}(z,z_0)e_1(\theta,\ell) + Q^{\ell+\frac{1}{2}}_{s,k}(z,z_0)e_2(\theta,\ell)\,,\end{aligned}$$

with

$$\begin{aligned}e_1(\theta,\ell) &:= e^{i(\ell-\frac{1}{2})\theta}\begin{pmatrix}1\\0\end{pmatrix}\,,\\ e_2(\theta,\ell) &:= e^{i(\ell+\frac{1}{2})\theta}\begin{pmatrix}0\\1\end{pmatrix}\,.\end{aligned}$$

The spherical functions associated to the Laplacian D_k are

$$P^{\ell}_{s,k}(z,a_\nu) := \frac{1}{\Gamma(1-\ell)}\, t^{-\ell/2}(1-t)^s F(s+k, s-k-\ell; 1-\ell; t)\,,$$

$$Q^{\ell}_{s,k}(z,a_\nu) := \frac{e^{i\ell\pi}}{4\pi}\frac{\Gamma(s-k)\,\Gamma(s+k+\ell)}{\Gamma(2s)}\, t^{-\ell/2}(1-t)^s F(s+k, s-k-\ell; 2s; 1-t)\,,$$

where

$$t = \frac{\cosh r - 1}{\cosh r + 1}\,,$$

and $F(a,b;c,z)$ is the hypergeometric function. The reader should note that although we have adopted the notation of Fay [2], we have altered slightly the definitions of $P^{\ell}_{s,k}(z,a_\nu)$ and $Q^{\ell}_{s,k}(z,a_\nu)$.

Furthermore, as in SMJ III, we can prove that the dimension of $W_k(L,A)$ is n, the number of singularities appearing in Definition 2.1, by constructing a *canonical basis* $\{\Psi_\mu\}$. If $\{\Psi_\mu\}$ is any basis of $W_k(L,A)$ and $c_{\mu\nu}(\ell,\pm)$ the local expansion coefficients of Ψ_μ at the point a_ν, we introduce the matrices

$$\begin{aligned}
[C_\pm(j)]_{\mu\nu} &:= c_{\mu\nu}(\ell_\nu \pm j, \pm)\,,\\
G &:= -(\cos\pi L)^{-1} C_-^{-1}(0) C_+(0)\,,\\
\Lambda^+ &:= C_+^{-1}(0) C_+(1)\,,\\
F^+ &:= [A,\Lambda^+] - (A-\tilde{A})(L + k + \tfrac{1}{2})\,,\\
\tilde{A} &:= G^{-1}\bar{A}G\,.
\end{aligned} \tag{2.4}$$

The canonical basis is characterized by the condition $C_+(0) = I$. In what follows we need to invert the matrix $(A-\tilde{A})$. We define

$$\mathcal{S}_1 := \Big\{(a_1,\ldots,a_n) \in \mathbf{H}\times\ldots\times\mathbf{H} \Big| \det(A-\tilde{A}) = 0\Big\}$$

and we will assume the existence of points $(a_1,\ldots,a_n)$ that are not in $\mathcal{S}_1$. Note that this is trivially true for *fixed* G, but that $G = G(A,\bar{A})$ as determined by the deformation equations below.

§3. Extended Equations. In any isomonodromy problem, a recurring idea, going back at least as far to Richard Fuchs [3], is to extend the given differential equation to a total system of differential equations; and then to demand integrability(see [15] for additional references to the early literature). As far as the authors are aware, the application of this idea to *partial* differential equations first appeared in SMJ III. The partial differential operators appearing in SMJ are the elliptic operators

$$\Delta_E := (\partial_x^2 + \partial_y^2) - m^2 id$$

and

$$\Gamma_E := \begin{pmatrix} m & -\partial \\ -\bar{\partial} & m \end{pmatrix}$$

acting on either functions or spinors, respectively, defined on $\mathbf{R^2}$. The extension was obtained by first introducing the collection of symmetry operators

$$\mathcal{F}_{E(2)} := \{\partial, \bar{\partial}, z\partial - \bar{z}\bar{\partial}\}$$

for Δ_E and the corresponding spinor version,

$$\mathcal{F}^S_{E(2)} = \left\{\partial \begin{pmatrix} 1 & 0 \\ 0 & 1 \end{pmatrix}, \bar{\partial} \begin{pmatrix} 1 & 0 \\ 0 & 1 \end{pmatrix}, (z\partial - \bar{z}\bar{\partial}) \begin{pmatrix} 1 & 0 \\ 0 & 1 \end{pmatrix} + \frac{1}{2} \begin{pmatrix} 1 & 0 \\ 0 & -1 \end{pmatrix}\right\},$$

for Γ_E. The point is that the vector fields in $\mathcal{F}_{E(2)}$ are the infinitesimal generators of the Euclidean group $E(2)$, $\mathbf{R^2} = E(2)/O(2)$, and Δ_E is the Casimir invariant (precisely, any differential operator commuting with all the vector fields in $\mathcal{F}_{E(2)}$ is a constant coefficient polynomial of Δ_E). As discussed in [25], if $\mathbf{R^2}$ is replaced by $\mathbf{H} = SL_2(\mathbf{R})/SO(2)$, Δ_E by the $SL_2(\mathbf{R})$-Casimir invariant $\Delta_{\mathbf{H}} := y^2(\partial_x^2 + \partial_y^2) - s(s-1)id$, $s > 1$, and $\mathcal{F}_{E(2)}$ by the infinitesimal generators of $SL_2(\mathbf{R})$, $\mathcal{F}_{SL_2(\mathbf{R})} := \{F_1, F_2, F_3\}$ where

$$\begin{aligned} F_1 &:= \partial + \bar{\partial}, \\ F_2 &:= z\partial + \bar{z}\bar{\partial}, \\ F_3 &:= z^2\partial + \bar{z}^2\bar{\partial}; \end{aligned}$$

then a corresponding extended system of equations for the isomonodromy problem associated to $\Delta_{\mathbf{H}} f = 0$ can be found in a manner paralleling SMJ. We now proceed with this idea as applied to the Dirac operator Γ_k.

The spinor version of $\mathcal{F}_{SL_2(\mathbf{R})}$ is $\mathcal{F}^S_{SL_2(\mathbf{R})} = \{F_1^S, F_2^S, F_3^S\}$ where

$$\begin{aligned} F_1^S &= (\partial + \bar{\partial}) \begin{pmatrix} 1 & 0 \\ 0 & 1 \end{pmatrix}, \\ F_2^S &= (z\partial + \bar{z}\bar{\partial}) \begin{pmatrix} 1 & 0 \\ 0 & 1 \end{pmatrix}, \\ F_3^S &= (z^2\partial + \bar{z}^2\bar{\partial}) \begin{pmatrix} 1 & 0 \\ 0 & 1 \end{pmatrix} + (z - \bar{z}) \begin{pmatrix} k+1 & 0 \\ 0 & k \end{pmatrix}. \end{aligned} \tag{3.1}$$

We have the commutation relations $[F_1^S, F_2^S] = F_1^S$, $[F_1^S, F_3^S] = 2F_2^S$, and $[F_2^S, F_3^S] = F_3^S$. It can be shown that any first order matrix differential operator commuting with the Dirac operator Γ_k is a linear combination of F_1^S, F_2^S, and F_3^S (and, of course, a multiple of the identity and a multiple of Γ_k). We write $p(\cdot,\cdot)$ for a constant coefficient polynomial in two variables, and define

$$W_k^1(L, A) := \{p(F_1^S, F_2^S)\Psi \,|\, \deg p \leq 1, \Psi \in W_k(L, A)\}.$$

We can show that $F_3^S \Psi_\mu \in W_k^1(L, A)$. Clearly, a basis for $W_k^1(L, A)$ is $\{F_1^S \Psi_\mu, F_2^S \Psi_\mu, \Psi_\mu\}$ where $\{\Psi_\mu\}$ is any basis for $W_k(L, A)$. Thus we have extended our system of equations $\Gamma_k \Psi_\mu = 0$ ($\mu = 1, \ldots, n$) to also include

$$F_3^S \Psi_\mu = \sum_{\lambda=1}^n \left\{B_{\mu\lambda}^{(1)} F_1^S \Psi_\lambda + B_{\mu\lambda}^{(2)} F_2^S \Psi_\lambda + B_{\mu\lambda}^{(3)} \Psi_\lambda\right\}, \ \mu = 1, \ldots, n; \tag{3.2}$$

for some coefficients $B^{(j)}_{\mu\lambda}$ ($j = 1,2,3; \mu, \lambda = 1, \ldots, n$) which are independent of z and $\bar{z}$. We will denote by $B^{(j)}$ the $n \times n$ matrix whose coefficients are $B^{(j)}_{\mu\lambda}$. As in SMJ, the matrices $B^{(j)}$ can be expressed in terms of the local expansion coefficient matrices $C_{\pm}(0)$ and $C_{\pm}(1)$ by equating coefficients in a local expansion of both sides of (3.2). The resulting formulas for the case of the canonical basis are

$$\begin{aligned} B^{(1)} &= -(A-\tilde{A})\tilde{A}(A-\tilde{A})^{-1}A \\ &= -(A-\tilde{A})A(A-\tilde{A})^{-1}\tilde{A}\,, \\ B^{(2)} &= (A^2-\tilde{A}^2)(A-\tilde{A})^{-1} \\ &= \tilde{A} + (A-\tilde{A})A(A-\tilde{A})^{-1}\,, \\ B^{(3)} &= -\big(F^+A + AF^+\big)(A-\tilde{A})^{-1} + \big(A^2-\tilde{A}^2\big)\big(A-\tilde{A}\big)^{-1}F^+(A-\tilde{A})^{-1}\,, \end{aligned}$$

where F^+ and $\tilde{A}$ are defined in (2.4).

We now introduce a representation of Ψ_μ so that the Dirac equation (2.1) is automatically satisfied. For the case $k = 0$ this is the hyperbolic Laplace transform (cf. [7]); that is, if

$$\psi(x,y) = \int_C P_s(x-\xi, y)\hat{\psi}(\xi)\,d\xi$$

where

$$P_s(x,y) = \Big(\frac{y}{x^2+y^2}\Big)^s,$$

then ψ satisfies $D_0\psi = s(s-1)\psi$ where $D_0 = y^2(\partial_x^2 + \partial_y^2)$. For $k \neq 0$ the components of Ψ satisfy (2.4). The generalization of the hyperbolic Laplace transform to spinors is

$$\Psi(x,y) = \int_C \mathcal{P}_{k,s}(x-\xi, y)\hat{\Psi}(\xi)\,d\xi \tag{3.3}$$

where

$$\mathcal{P}_{k,s}(x,y) = P_s(x,y)\begin{pmatrix} \big(\frac{x-iy}{x+iy}\big)^{k+1} & 0 \\ 0 & \big(\frac{x-iy}{x+iy}\big)^{k} \end{pmatrix}.$$

One way to discover this kernel is to write an integral representation for the basic spherical functions $P^{\ell}_{s,k}(z, z_0)$ (cf. [2]). A formal derivation follows from the fact that

$$D_k\Big(\frac{x-iy}{x+iy}\Big)^k P_s(x,y) = s(s-1)\Big(\frac{x-iy}{x+iy}\Big)^k P_s(x,y)\,.$$

The following proposition, whose proof is similar to Proposition 2.1 in [25], tells us how the action of the vector fields F_j^S translates into operators H_j acting on $\hat{\Psi}(\xi)$.

Proposition 3.1. *Let F_j^S be the vector fields defined by (3.1), and let Ψ and $\hat{\Psi}$ be related by (3.3), then*

$$(F_j^S\Psi)(x,y) = \int_C \mathcal{P}_{k,s}(x-\xi, y)(H_j\hat{\Psi})(\xi)\,d\xi$$

where

$$H_1 = \partial_\xi\, I$$
$$H_2 = (1 - s + \xi\partial_\xi)\, I$$
$$H_3 = \big(2(1-s)\xi + \xi^2\partial_\xi\big)\, I$$

where I is the 2×2 identity matrix.

We now show that the transform of the canonical basis satisfies an ordinary differential equation of Fuchsian type:

Theorem 3.2. *Let* $\left\{\Psi_\mu = \begin{pmatrix}\psi_{1,\mu}\\ \psi_{2,\mu}\end{pmatrix}\right\}$ *denote the canonical basis for $W_k(L,A)$, let*

$$\Psi_\mu(z,\bar z) = \int_{\mathcal{C}} \mathcal{P}_{k,s}(x-\xi,y)\hat\psi_\mu(\xi)\, d\xi\,, \tag{3.4}$$

the contour $\mathcal{C}$ encircles a_ν and $\bar a_\nu$ ($\nu = 1,\ldots,n$) with z and $\bar z$ on the outside, and $\hat\Psi_j$ ($j=1,2$) the $n\times 1$ column vector $\begin{pmatrix}\hat\psi_{j,1}\\ \vdots\\ \hat\psi_{j,n}\end{pmatrix}$. *Then $\hat\Psi_j$ ($j=1,2$) satisfies the Fuchsian system*

$$\frac{d\hat\Psi_j}{d\xi} = \sum_{\nu=1}^{n}\Big(\frac{A_\nu}{\xi - a_\nu} + \frac{A'_\nu}{\xi - \bar a_\nu}\Big)\hat\Psi_j \qquad (j=1,2) \tag{3.5}$$

where A_ν and A'_ν are rank one $n\times n$ matrices given by

$$\begin{aligned} A_\nu &= E_\nu\big(A-\tilde A\big)^{-1}\big[(1-s)B^{(2)} + B^{(3)} - 2(1-s)a_\nu I\big]\\ A'_\nu &= -\tilde E_\nu\big(A-\tilde A\big)^{-1}\big[(1-s)B^{(2)} + B^{(3)} - 2(1-s)\bar a_\nu I\big]\end{aligned} \tag{3.6}$$

and E_ν is the diagonal matrix with all zeros along the main diagonal excepts for a 1 at the νth position, and $\tilde E_\nu = G^{-1}E_\nu G$. Furthermore, the matrices A_ν and A'_ν satisfy the conditions

$$\begin{aligned} \sum_{\nu=1}^{n}(A_\nu + A'_\nu) &= 2(s-1)I\\ Tr(A_\nu) &= s + \ell_\nu + k - \frac12\\ Tr(A'_\nu) &= s - \ell_\nu - k - \frac32 \end{aligned} \tag{3.7}$$

Sketch of Proof: We use (3.3) and Proposition 3.1 to transform the system (3.2) to the system of ordinary differential equations:

$$\Delta(\xi)\frac{d\hat\Psi_j}{d\xi} = \big[(1-s)B^{(2)} + B^{(3)} - 2(1-s)\xi I\big]\hat\Psi_j$$

where $\Delta(\xi) = \xi^2 I - \xi B^{(2)} - B^{(1)}$. From the computation

$$\begin{aligned}
&(A - \tilde{A})\big[(\xi I - A)^{-1} - (\xi I - \tilde{A})^{-1}\big]^{-1} \\
&\quad = (A - \tilde{A})\Big[(\xi I - \tilde{A})^{-1}\big[(\xi I - \tilde{A}) - (\xi - A)\big](\xi I - A)^{-1}\Big]^{-1} \\
&\quad = (A - \tilde{A})\big[(\xi I - A)(A - \tilde{A})^{-1}(\xi I - \tilde{A})\big] \\
&\quad = \xi^2 I - (A - \tilde{A})A(A - \tilde{A})^{-1}\xi - \tilde{A}\xi + (A - \tilde{A})A(A - \tilde{A})^{-1}\tilde{A} \\
&\quad = \xi^2 I - \xi B^{(2)} - B^{(1)}
\end{aligned}$$

it follows that

$$\Delta^{-1}(\xi) = \big[(\xi I - A)^{-1} - (\xi I - \tilde{A})^{-1}\big](A - \tilde{A})^{-1}$$

Thus the transformed system has the form

$$\frac{d\hat{\Psi}_j}{d\xi} = A(\xi)\hat{\Psi}_j$$

where

$$A(\xi) = \big[(\xi I - A)^{-1} - (\xi I - \tilde{A})^{-1}\big](A - \tilde{A})^{-1}\big[(1 - s)B^{(2)} + B^{(3)} - 2(1 - s)\xi I\big]$$

The advantage of this last representation for $A(\xi)$ is that the analytic structure in ξ can be readily seen and hence it is straightforward to calculate a partial fraction decomposition:

$$A(\xi) = \sum_{\nu=1}^{n} \frac{A_\nu}{\xi - a_\nu} + \frac{A'_\nu}{\xi - \bar{a}_\nu}$$

where A_ν and A'_ν are given by (3.6). It is now clear that the sum condition on A_ν and A'_ν and the rank one property are satisfied. The trace conditions in (3.7) follow from some technical identities obtained from local expansions (see [25]).

We turn to the choice of the contour $\mathcal{C}$. First of all, $\hat{\Psi}(\xi)$ is chosen to be the unique (up to an overall constant) solution to (3.5) that has the local exponents given by the trace conditions (3.7). This is possible since $n-1$ exponents at a point a_ν ($\bar{a}_\nu$) are zero because of the rank one property and the remaining nonzero exponent is given by $\mathrm{Tr}(A_\nu)$ ($\mathrm{Tr}(A'_\nu)$). We claim the contour encircling a_ν and $\bar{a}_\nu$ ($\nu = 1, \ldots, n$) with z and $\bar{z}$ on the outside is such a contour (see discussion in [25]).

From Theorem 5.1 it follows that the *local* monodromy of (3.5) does not change when the a_ν and $\bar{a}_\nu$ are varied. Since (3.5) is a Fuchsian system (the sum condition in (3.7) means that ∞ is a regular singular point), the necessary and sufficient conditions for isomonodromy are that A_ν and A'_ν satisfy the Schlesinger equations (the Schlesinger equations (1.1) have to be slightly modified since it was assumed that ∞ was a regular point). Thus it is consistent with Theorem 5.1 to require isomonodromy of the transformed equations (3.5).

Note Added: For the hyperbolic Laplacian case ($k = 0$), the deformation problem has been studied by Narayanan and Tracy (to appear in Nuclear Physics B[FS]) as a function of the curvature. It is shown that in the zero curvature limit the hyperbolic plane results connect to the Euclidean space results of Sato, Miwa, and Jimbo.

Acknowledgments. The first author would like to thank Professor J. Keizer for the invitation to visit the Institute of Theoretical Dynamics where part of this work was done. The second author would like to thank Professor M. Berger for the invitation to speak at the January, 1989 AMS meeting in Phoenix where the $k = 0$ case of this note was presented.

REFERENCES

1. R. Davey, *SMJ Analysis of Monodromy Fields*, thesis, unpublished (Univ. Arizona, 1988).
2. J. Fay, *Fourier coefficients of the resolvent for a Fuchsian group*, J. Reine Angew. Math. **293** (1977), 143–203.
3. R. Fuchs, *Über lineare homogene differentialgleichungen zweiter ordnung mit drei im endlichen gelegene wesentlich singulären stellen*, Math. Annalen **63** (1907), 301–321.
4. R. Garnier, *Sur des équations différentielles du troisième ordre dont l'intégrale générale est uniforme et sur une classe d'équations nouvelles d'ordre supérieur dont l'intégrale générale a ses points critques fixes*, Ann. Sci. Ecole Norm. Sup. (3) **29** (1912), 1–126.
5. R. Gerard, *La geometrie des transcendantes de P. Painlevé*, in Mathématique et Physique. Séminaire de l'Ecole Normale Supérieure 1979–1982, L. B. de Monvel, A. Douady, and J.-L. Verdier, eds. (Birkhäuser, Boston, 1983), 323–352.
6. M. B. Green, J. H. Schwarz, and E. Witten, Superstring Theory, Vol. 2 (Cambridge Univ. Press, Cambridge, 1987).
7. S. Helgason, Groups and Geometric Analysis: Integral Geometry, Invariant Differential Operators, and Spherical Functions (Academic Press, Orlando, 1984).
8. E. L. Ince, Ordinary Differential Equations (Dover Publ., N. Y., 1956).
9. N. M. Katz, *An overview of Deligne's work on Hilbert's Twenty-First Problem*, in Mathematical Developments Arising from Hilbert Problems (Amer. Math. Soc., Providence, 1976), 537–557.
10. G. E. Latta, *The solution of a class of integral equations*, J. Rational Mech. Anal. **5** (1956), 821–833.
11. B. Malgrange, *Sur les deformations isomonodromiques. I. Singularites regulieres*, in Mathématique et Physique. Séminaire de l'Ecole Normale Supérieure 1979–1982, L. B. de Monvel, A. Douady, and J.-L. Verdier, eds. (Birkhäuser, Boston, 1983), 401–426.
12. B. M. McCoy, C. A. Tracy, and T. T. Wu, *Painlevé functions of the third kind*, J. Math. Phys. **18** (1977), 1058–1092.
13. B. M. McCoy and T. T. Wu, The Two-Dimensional Ising Model (Harvard Univ. Press, Cambridge, 1973).
14. J. Myers, *Symmetry in scattering by a strip*, thesis, unpublished (Harvard Univ., 1962). *Wave scattering and the geometry of a strip*, J. Math. Phys. **6** (1965), 1839–1846.
15. K. Okamoto, *Isomonodromic deformation and Painlevé equations, and the Garnier system*, J. Fac. Sci. Univ. Tokyo **33** (1986), 575–618.
16. J. Palmer, *Monodromy fields on* $\mathbf{Z}^2$, Commun. Math. Phys. **102** (1985), 175–206.
17. J. Palmer, *Pfaffian bundles and the Ising model*, Commun. Math. Phys. **120** (1989), 547–574.
18. J. Palmer, *Determinants of Cauchy-Riemann operators as τ-functions*, preprint.
19. J. Palmer and C. Tracy, *Two-dimensional Ising correlations: The SMJ analysis*, Adv. in Appl. Math. **4** (1983), 46–102.
20. D. Quillen, *Determinants of Cauchy-Riemann operators on a Riemann surface*, Funct. Anal. Appl. **19** (1985), 37–41.

21. H. Röhrl, *Das Riemann-Hilbertsche problem der theorie der linearen differentialgleichungen*, Math. Ann. **133** (1957), 1–25.

22. M. Sato, T. Miwa, M. Jimbo, *Holonomic quantum fields I–V*. Publ. Res. Inst. Math. Sci., Kyoto Univ. **14** (1978), 223–267; **15** (1979), 201–278; **15** (1979), 577–629; **15** (1979), 871–972; **16** (1980), 531–584.

23. L. Schlesinger, *Über eine klasse von differentialsystemen beliebiger ordnung mit festen kritischen punkten*, J. Reine Angew. Math. **141** (1912), 96–145.

24. G. Segal and G. Wilson, *Loop groups and equations of KdV type*, Publ. Math. I. H. E. S. **61** (1985), 5–65.

25. C. A. Tracy, *Monodromy preserving deformation theory of the Klein-Gordon equation in the hyperbolic plane*, Physica **D34** (1989), 347–365.

26. E. Witten, *Quantum field theory, Grassmannians, and algebraic curves*, Commun. Math. Phys. **113** (1988), 529–600.

27. T. T. Wu, B. M. McCoy, C. A. Tracy, and E. Barouch, *Spin-spin correlation functions for the two-dimensional Ising model: Exact theory in the scaling region*, Phys. Rev. **B13** (1976), 316–374.

JOHN PALMER
DEPARTMENT OF MATHEMATICS
UNIVERSITY OF ARIZONA
TUCSON, AZ 85721

CRAIG A. TRACY
DEPARTMENT OF MATHEMATICS
AND
INSTITUTE OF THEORETICAL DYNAMICS
UNIVERSITY OF CALIFORNIA
DAVIS, CA 95616

Contemporary Mathematics
Volume **108**, 1990

Bifurcation from Equilibria for Certain Infinite-Dimensional Dynamical Systems

M. S. Berger* and M. Schechter*

Nonlinear parabolic partial differential equations exhibit a broad class of bifurcation phenomena. Indeed, every kind of bifurcation known for first order systems of nonlinear ordinary differential equations has an analogue for parabolic systems when spatial variables are included. Thus, many of the problems for parabolic partial differential equations, with appropriate boundary conditions, exhibit Hofp bifurcation, chaotic dynamics, saddle-node bifurcation, to mention only a few types of nonlinear phenomena of bifurcation. The whole idea of center manifolds in bifurcation phenomena for dynamical systems is connected with this idea.

In the following article, we adopt a different viewpoint. We look at the whole problem from a functional analysis point of view. We regard the parabolic system as arising from an elliptic problem by adding time dependents. We inquire if it is possible to use the detailed analysis of the resulting elliptic system to study the time-dependent situation. Moreover, we focus on special classes of solutions of the time-dependent problem, namely, periodic solutions. In so doing, we avoid studying the initial value problem for the parabolic system in question and we thus avoid chaotic dynamics entirely.

1980 *Mathematics Subject Classification* (1985 *Revision*). 49F22, 53A10.
The final (detailed) version of this paper will be submitted for publication elsewhere.
*Research was partially supported by grants from the AFOSR and the NSF.

A new feature of our viewpoint is an attempt to use the new infinite-dimensional theory of singularities of maps initiated by Berger and Church in articles mentioned in the bibliography for dynamical systems defined by parabolic partial differential equations. In this context, we regard a nonlinear parabolic partial differential equation subject to appropriate boundary conditions as a mapping between two infinite-dimensional spaces. The linearization of this mapping is assumed to be sufficiently well-defined so that it is possible to distinguish between regular and singular points. For this reason, we focus attention on periodic time-dependents. This assumption enables us to assume the mapping in question B is a Fredholm operator of index 0. Once this fact has been established, we are able to ue the infinite-dimensional theory of singularities to determine very explicit types of bifurcation at a singular point. In this article, we consider only bifurcation of the Whitney fold type. However, in theory, there is no reason why any singularities of codimension 1 could not be considered.

Introduction

In this article we consider the bifurcation of T-periodic solutions of the parabolic boundary-value problem

$$(1) \qquad \begin{cases} \dfrac{\partial u}{\partial t} + Lu + f(x, u) = g(x, t) & \text{in } \Omega \\ u|_{\partial\Omega} = 0 \end{cases}$$

obtained for a given T-periodic smooth forcing term $g(x, t)$ in (1). Here $x \in \Omega$ an arbitrary bounded domain in $\mathbb{R}^N$ with boundary $\partial\Omega$. Moreover, L is a uniformly elliptic second order formally self-adjoint second order differential operator defined on Ω. The smooth function $f(u, x) \in \mathbb{R}^1 \times \Omega$ is specialized for the present to be convex with f_u bounded and to have the asymptotic properties

$$\lambda_2 > \lim_{t\to\infty} \frac{f(t, x)}{t} > \lambda_1 \quad ; \quad \lambda_1 > \lim_{t\to-\infty} \frac{f(t, x)}{t} > 0$$

where λ_1 and λ_2 are the lowest two eigenvalues of L relative to Ω.

The bifurcation phenomena reported here differs substantially from the Hopf bifurcation [1] generally discussed for the change from equilibria to T-periodic solutions. It represents an extension to time-dependent from steady problems of the bifurcation mathematical structures recently discussed by me relative to infinite-dimensional Whitney fold and cusp singularities [2].

§1. The Static Case

Consider the special case, in which the forcing function g in (1) does not depend on t. Then the periodic solutions of (1) also do not depend on t, so (1) reduces to the nonlinear elliptic Dirichlet problem

$$Lu + f(x, u) = g(x) \quad \text{in } \Omega \tag{2}$$

$$u|_{\partial\Omega} = 0$$

This problem has been studied extensively, (cf. [3]).

In terms of functional analysis, it suffices to consider weak solutions of (2) (for $g \in L_2(\Omega)$) in the Sobolev space $H = \dot{W}_{1,2}(\Omega)$. Then the bifurcation points of (2) coincide with the singular points of the operator $A : Lu + f(x, u) : H \to H$ defined via duality, (cf. [3]).

This operator A turns out to be a nonlinear Fredholm operator of index zero whose singular points have the following two properties: (cf. [2])

(a) The singular points of A, $S(A)$ form a connected infinite-dimensional manifold M. In fact M is a hypersurface in H of codimension one.

(b) All the singular points of A are infinite-dimensional folds (in the appropriate infinite-dimensional generalization of Whitney ([2]).

Thus one is lead to inquire: What happens to these results for the time-dependent case (1)?

§2. The Time-Dependent Periodic Case

We now consider the full equation (1) with the forcing term $g = g(x, t)$ depending explicitly on t in a T-periodic manner. The operator

$$(3) \qquad Bu = \frac{\partial u}{\partial t} + Lu = f(x, u) \qquad u|_{\partial\Omega} = 0$$

comprising the left hand side of (2) can be regarded as a smooth mapping between real Hilbert spaces X and Y of T-periodic function in t with

$$X = W_{1,2}[(0, T), H]$$

and

$$Y = L_2[(0, T), H]$$

with the usual norms.

In these terms we can prove the following extensions of our discussion of §1.

Theorem 1 The operator B regarded as a C^1 mapping betweeen the real Hilbert spaces X and Y is a nonlinear Fredholm operator of index zero.

Theorem 2 Regular points of the mapping A are regular points of B. Moreover singular points of A are singular points of B and, in fact, for any $u \in H$

$$(4) \qquad \dim\ker A'(u) = \dim\ker B'(u)$$

Theorem 3 Any $u \in H \cap S(A)$ is an (infinite-dimensional) Whitney fold for B, provided B is regarded as a mapping between X and Y.

Theorem 4 For $g(x, t)$ smooth, restricted to a small neighborhood of a singular value of A in Y, equation (1) has exactly 2, 1 or 0 smooth, real T-periodic solutions in an appropriate neighborhood of the associated singular point in X.

In other words, bifurcation from equilibrium for T-periodic solutions of (1) occurs precisely at singular points u at A. Moreover the fold-type bifurcations occuring for the static problem go over to the same fold bifurcations for the periodic problem.

§3. Sketch of the Proofs

On the Proof of Theorem 1 The operator Bu acting between X and Y can be written as the sum of the invertible linear operator $B_0 = u_t + Lu$ plus a compact operator, as follows from the Sobolev-Kondrachov results. Here

$$(5) \qquad B_0^{-1} w = e^{-tL}(I - e^{-TL})^{-1} Kw(T) + Kw(t)$$

$$\text{where} \quad Kw(t) = \int_0^t e^{(s-t)L} w(s)ds$$

On the Proof of Theorem 2 Suppose (based on Theorem 1) that for $v(x, t) \in \text{Ker } B'(u)$

$$(6) \qquad v(x, t) = \sum_{i=1}^{v} C_k(t)v_k(x) + w(x, t)$$

where $w \perp v_k$ for all t, k and $v_1, v_2 \ldots v_v$ form an orthonormal basis for $\text{Ker } A'(u)$. We prove C_k is independent of t and $w(x, t) \equiv 0$. The result establishes equation (4), and so the full result.

On the Proof of Theorem 3 This follows from our characterization of an infinite-dimensional Whitney fold between X and Y described above. This requires, for example, $\dim\ker A'(u) = 1$ as established above for any singular point u of A. So that via (3) $\dim\ker B'(u) = 1$. The second requirement for a fold follows from the properties of f described at the beginning of this note, especially the convexity of f in u.

On the Proof of Theorem 4 A consequence of parabolic regularity theory for this problem and the canonical normal form for an infinite-dimensional fold singularity [2].

§4. Extensions

The analysis given here can be directly extended to more equations and systems than (1). For example L can be of order $2m$ and $f(u)$ need not be bounded. The associated bifurcation or convex analysis must be considerably extended to cover Whitney cusps as well as more general singularities (cf. [2]).

Bibliography

1. J. Marsden and M. McCracken, Hopf Bifurcation, Springer (1979).
2. M. S. Berger, P. T. Church and J. Timourian, Folds and Cusp in a Banach Space with Application I, Indiana Math Journal, Vol. 34, (1985), pp. 1-19.
3. M. S. Berger, Nonlinearity and Functional Analysis, Academic Press, New York (1977).

University of Massachusetts
Amherst, MA 01003
and
University of California
Irvine, CA 92717

Contemporary Mathematics
Volume **108**, 1990

On Dynamics of Discrete and Continuous σ-models (Chiral Fields) with Values in Riemannian Manifolds

VICTOR SHUBOV

1. We describe below the results of **[1]** on the problem of unique global (in time) solvability of some class of infinite systems of nonlinear ordinary differential equations describing functions with values in Riemannian manifolds. These equations are, on the one hand, discrete analogues of equations of relativistic σ-models (hyperbolic harmonic maps), and, on the other hand, they can be interpreted as equations of motion for a special class of infinite models of classical statistical mechanics. We describe these equations using the language of statistical mechanics.

A model of the class mentioned above is a model of an infinite crystal—an infinite system of interacting anharmonic oscillators. We consider the case when the configuration space of each oscillator is a Riemannian manifold; in other words, these oscillators are constrained. The equations of motion of the above model can be obtained by taking the variation of the formal action functional

$$S(\hat{q}) = \int_{-\infty}^{\infty} \left[\sum_{k\in\Gamma} \frac{1}{2} |\dot{q}_k(t)|_g^2 - U(\hat{q}(t)) \right] dt. \tag{1}$$

We use the following notations: The index k runs over the set of all vertexes of an infinite connected graph Γ; Γ is a model of crystal lattice. $q_k(t)$ is the configuration of the kth oscillator at the moment t. The functions q_k take values in a smooth manifold Q, equipped with a Riemannian metric g; Q is a configuration space of each oscillator. $\dot{q}_k(t)$ is the derivative of q_k with respect to t—the velocity of the kth oscillator—the vector tangent to the curve q_k at the point $q_k(t)$. $|\dot{q}_k(t)|_g$ designates the norm of this vector

---

1980 *Mathematics Subject Classification* (1985 *Revision*). Primary 82A05, 53C21; Secondary 35L15, 58G16.

This paper is in final form and no version of it will be submitted for publication elsewhere.

in the metric g; in local coordinates, $|\dot{q}_k(t)|_g^2 = g_{\alpha\beta}(q_k(t))\dot{q}_k^\alpha(t)\dot{q}_k^\beta(t)$. The first term in the brackets in (1) is the kinetic energy of the system. U is the potential energy—a function on the infinite dimensional configuration space Q^Γ of the system. $\hat{q}(t)$ is the configuration of the whole infinite system. The conditions that U must satisfy will be given below.

The equations of motion defined by the action (1) in a local coordinate notations are

$$\ddot{q}_k^\alpha + \Gamma^\alpha_{\beta\gamma}(q_k)\dot{q}_k^\beta \dot{q}_k^\gamma = F_k^\alpha(\hat{q}); \qquad \alpha = 1, \dots, n, \qquad k \in \Gamma. \tag{2}$$

Here $\Gamma^\alpha_{\beta\gamma}$ are the Christoffel coefficients corresponding to the metric g; $\dim Q = n$. $F_k(\hat{q})$ is the force acting on the kth particle in a given configuration of the system $F_k^\alpha(\hat{q}) = g^{\alpha\beta}\partial U/\partial q_k^\beta$. The invariant form of (2) is

$$\nabla_{\dot{q}_k} \dot{q}_k = F_k(\hat{q}), \qquad k \in \Gamma, \tag{3}$$

where ∇ is the Levi–Civita connection of the metric g.

Fix the initial conditions

$$q_k(0) = q_{0k} \in Q, \qquad \dot{q}_k(0) = v_{0k} \in TQ_{q_{0k}}, \qquad k \in \Gamma. \tag{4}$$

(TQ_q is a tangent space of Q at the point $q \in Q$.)

The main result of **[1]** consists of the following: We describe a class of Riemannian manifolds (Q, g), which we call a class of admissible constraints; we also describe some class of potential energy functionals U, which we call a class of admissible interactions. For these constraints and interactions we prove

THEOREM 1. *If (Q, g) belongs to the class of admissible constraints and U belongs to the class of admissible interactions, then the Cauchy problem* (2), (4) *is uniquely globally (in t) solvable for all initial data* (4).

2. Before describing the above classes of constraints and interactions we give some commentary.

(a) The proof of the existence and the uniqueness theorems for the equations of motion of an infinite system is the first step in a rigorous approach to nonequilibrium statistical mechanics. In other words, the problem is to construct dynamics of an infinite system. The main obstacle to the solution of this problem consists in the following: One cannot usually prove the existence and the uniqueness for all initial data; it can be done only for some classes of initial data. So, the dynamics of an infinite system can be usually defined only on some subset of the phase space. The problem is to make this subset sufficiently large. For the purpose of the nonequilibrium statistical mechanics, this subset is sufficient to be of full measure with respect to any Gibbs state. The problem of construction of dynamics is solved now only for some model systems; for real systems it is unsolved.

The above result shows that we can present a class of models that have dynamics defined on the whole infinite dimensional phase space of the system. We must emphasize that this result is not valid for arbitrary Riemannian manifolds. In a general case, one has to introduce limitations on the initial velocities.

(b) Now we recall the results of **[2]**. In **[2]** the problem of global existence of hyperbolic harmonic maps from $\mathbf{R}^2$ to an arbitrary complete Riemannian manifold (Q, g) was considered. The equation for such maps $\varphi: \mathbf{R}^2 \to Q: (x, t) \mapsto \varphi(x, t)$ can be obtained by taking the variation of the functional

(5) $$S(\varphi) = \int_{\mathbf{R}^2} (|\varphi_t|_g^2 - |\varphi_x|_g^2)\, dx\, dt.$$

Here φ_t, φ_x are the partial derivatives of φ. The corresponding equation in local coordinates has the form

(6) $$\varphi_{tt}^\alpha - \varphi_{xx}^\alpha + \Gamma_{\beta\gamma}^\alpha(\varphi)(\varphi_t^\beta \varphi_t^\gamma - \varphi_x^\beta \varphi_x^\gamma) = 0, \qquad \alpha = 1, \dots, n.$$

Fix the initial data

(7) $$\varphi(x, 0) = \varphi_0(x) \in Q, \qquad \varphi_t(x, 0) = \psi_0(x) \in TQ_{\varphi_0(x)}.$$

The main result of **[2]** consists of the following:

THEOREM 2. *Let Q be a C^s manifold $(s \geq 2)$, endowed with the complete Riemannian metric g of class C^s. Assume that the initial data (7) have the smoothness $\varphi_0 \in W_{2,\mathrm{loc}}^s$, $\psi_0 \in W_{2,\mathrm{loc}}^{s-1}$. Then there exists a unique solution φ of the Cauchy problem* (6), (7) *defined for all $t \in \mathbf{R}$ and having the following smoothness*: $\varphi \in C(\mathbf{R}, W_{2,\mathrm{loc}}^s)$, $\varphi_t \in C(\mathbf{R}, W_{2,\mathrm{loc}}^{s-1})$, $\varphi_{tt} \in C(\mathbf{R}, W_{2,\mathrm{loc}}^{s-2})$.

This theorem asserts that from the condition of geodesic completeness of (Q, g) (i.e., of unbounded extendability of the geodesics) there follows a significantly stronger fact: the unbounded extendability of the solutions of (6), (7). The result is valid for any (sufficiently regular) initial data. (The first-time derivatives of the solution at initial moment may increase at spatial infinity.) This is due to the hyperbolicity of (6).

If one considers the finite difference approximation (in x) of (5), (6) in the case of an arbitrary complete Riemannian manifold (Q, g), this effect of hyperbolicity will be lost. To prove the unique solvability for this approximation one has to impose the condition of uniform boundness of the initial velocities. Note now that the above approximation is a particular case of the model (1), (2). Theorem 1 shows that this effect of hyperbolicity is not lost in a special case, when (Q, g) belongs to the class of admissible constraints. The results of **[2]** were obtained independently of **[3]**.

3. Now we describe the class of admissible Riemannian manifolds (Q, g). We need to recall some facts from the Riemannian geometry.

Let (M, w) be an arbitrary complete Riemannian manifold. Let $\overset{\circ}{\nabla}$ be the Levi–Civita connection of the metric w. Let $A(M)$ be a class of linear connections on M defined by the following properties:

(i) Each $\nabla \in A(M)$ has the same geodesics as $\overset{\circ}{\nabla}$;
(ii) $\nabla w = 0$ for each $\nabla \in A(M)$.

It can be easily shown that there exists a natural one-to-one correspondence between $A(M)$ and the space $\Omega^3(M)$ of all differential 3-forms on M.

Now we define the above class of the manifolds (Q, g) in three steps.

Step 1. Let (M, w) be a complete Riemannian manifold, which has the following property: The class $A(M)$ contains at least one flat connection.

Step 2. Let N be a totally geodesic submanifold of M. It is natural to denote the induced metric on N by the same letter w.

Step 3. Let (Q, g) be a complete Riemannian manifold locally isometric to (N, w). It means that there exists a Riemannian manifold $(\widetilde{Q}, \widetilde{g})$ and two Riemannian coverings: $\tau: (\widetilde{Q}, \widetilde{g}) \to (Q, g)$, $\pi: (\widetilde{Q}, \widetilde{g}) \to (N, w)$.

Here is an example of a manifold (M, w) that satisfies the condition of Step 1. Let M be a Lie group and w a left and right invariant Riemannian metric on it. In this case $A(M)$ contains two flat connections satisfying Conditions (i), (ii). These are connections induced by the left and right actions of M on itself. In this case the class of (Q, g) consists of all symmetric and locally symmetric spaces of compact-Euclidean type.

4. The conditions on U are the conditions of sufficiently fast decrease of the interaction of two oscillators when the distance between these oscillators in the crystal lattice Γ tends to infinity. We will formulate here only the consequences of these conditions for the forces F_k.

(a) There exists a constant $C > 0$, such that

$$|F_k(\hat{q})|_g \leq C \quad \text{for all } k \in \Gamma, \quad \hat{q} \in Q^\Gamma.$$

(b) For any $k \in \Gamma$, $\hat{q}, \hat{q}' \in Q^\Gamma$,

$$|F_k(\hat{q}) - F_k(\hat{q}')|_g \leq \lim_{N\to\infty} \sum_{m \in B_k(N)} c_{km} \rho(q_m, q'_m),$$

where the constants c_{km} satisfy the condition

$$\lim_{N\to\infty} \sum_{m \in B_k(N)} c_{km} \leq C_1 < \infty.$$

Here we use the following notations: $B_k(N)$ is a "ball" in Γ with the center k and radius $N: B_k(N) = \{m \in \Gamma: \text{dist}(k, m) \leq N\}$, where $\text{dist}(k, m)$ is the number of edges of the minimal path of the graph Γ, connecting k and m.

5. In conclusion we mention that in **[4]** the existence of a weak solution of the BBGKY hierarchy corresponding to the model (1), (2) is proved. This hierarchy describes the evolution of a probability measure on the phase space of the system.

[5] contains a free-coordinate proof of the unique global solvability for the problem (2), (4) in the case of an arbitrary complete Riemannian manifold (Q, g). In this proof we assume that $|\dot{q}_k(0)|_g \leq C_0$ for all $k \in \Gamma$, and in the case when Q is noncompact, its sectional curvature K and the gradient of K are bounded.

REFERENCES

1. V. I. Shubov, *The dynamics of infinite classical anharmonic systems with constraints*, Proc. Steklov Inst. Math. **179** (1989), 203–224.
2. O. A. Ladyzhenskaya and V. I. Shubov, *Unique solvability of the Cauchy problem for the equations of the two-dimensional relativistic chiral fields taking values in complete Riemannian manifolds*, J. Sov. Math. **25** (1984), 855–864.
3. C.-H. Gu, *On the Cauchy problem for harmonic maps defined on two-dimensional Minkowsky space*, Comm. Pure Appl. Math. **33** (1980), 727–737.
4. V. I. Shubov, *Existence of a weak solution of the BBGKY hierarchy for infinite classical anharmonic systems with constraints*, J. Sov. Math. **40** (1988), 690–700.
5. V. I. Shubov, *Unique solvability of the Cauchy problem for the equations of discrete chiral fields with values in Riemannian manifolds*, J. Sov. Math. **30** (1985), 2353–2368.

TEXAS TECH UNIVERSITY

Contemporary Mathematics
Volume 108, 1990

DIRECT STUDY FOR SOME NONLINEAR ELLIPTIC CONTROL PROBLEMS

Srdjan Stojanovic [1]

INTRODUCTION. We consider two related optimal control problems. First, we consider an optimal control problem in which the state of the system is defined as the unique solution of an elliptic equation and where control is the zero order coefficient of the differential operator. In the second problem we consider the same state euqation but this time, control, i.e., the zero order coefficient of the differential operator is to be chosen implicitly, as a solution of another elliptic equation.

In both cases, after proving the existence of an optimal control, and deriving necessary conditions, i.e., an optimality system, we study this optimality system in detail, by providing a constructive existence theorem.

When constructing a solution, the difficulty is that uniform estimates, i.e. compactness, i.e., convergence for a subsequence of some natural iteration is not enough to solve the system. Hence, one has to invent a procedure with some monotonicity property, which in turn would guarantee the convergence. In this scheme uniqueness plays a fundamental role.

1. OPTIMAL DAMPING CONTROL. Let Ω be a bounded domain of R^n with $C^{1,1}$ boundary. Let

(1.1) $\quad f \in L^1(\Omega), \ q \geq 2, \ q > \frac{n}{2},$

and

(1.2) $\quad \lambda \in R_+ = \{\beta \mid \beta \in R, \ \beta \geq 0\}$

be given. For any $1 \geq 1$, set

(1.3) $\quad L^1_+(\Omega) = c \mid c \in L^1(\Omega), \ c \geq 0 \ \text{ a.e. in } \Omega\}.$

1980 Mathematics Subject Classification (1985 Revision). 49A22, 35J60.

[1]This research has been sponsored by DARPA under contract #1868-A037-A1 with Martin Marietta Energy Systems, Inc., which is under contract #DE-AC05-840R21400 with the U.S. Department of Energy.

The detailed version of this paper has been submitted for publication elsewhere.

For any $c \in L^2_+(\Omega)$, we define $y = y(c)$ as a solution of

$$(1.4)\qquad \begin{aligned} &-\Delta y + \lambda y + cy = f \quad \text{a.e. in } \Omega, \\ &y \in W^{2,2}(\Omega) \cap H^1_0(\Omega) \cap L^\infty(\Omega). \end{aligned}$$

REMARK 1.1. All results of this paper can be formulated for more general elliptic operators instead of $-\Delta$.

We observe that

$$(1.5)\qquad \begin{cases} \|y(c)\|_{L^\infty(\Omega)} \leq \text{const.} \\ \text{where const. does not depend on } c \in L^2_+(\Omega) \text{ and } \lambda \in R_+. \end{cases}$$

Next, for (q as in (1.1))

$$(1.6)\qquad y_d \in L^q(\Omega)$$

and $N > 0$ given, we define the cost functional J by

$$(1.7)\qquad J(c) = \frac{1}{2}\int_\Omega (y(c) - y_d)^2 dx + \frac{N}{2}\int_\Omega (c)^2 dx,$$

and we ask the following question:

(1.8) How do we find the optimal control?

i.e., to find the (damping) control $c_0 \in L^2_+(\Omega)$ such that

$$(1.9)\qquad J(c_0) = \min_{c \in L^2_+(\Omega)} J(c).$$

Under certain assumptions, we shall give a fairly complete answer to the question (1.8).

To start with, it is not difficult to see, using simple compactness and semicontinuity arguments, that an optimal control always exists.

PROPOSITION 1.1. <u>There exists an optimal control</u>.

The next theorem provides a characterization of the optimal control. Two results are given. The first result holds under general conditions already introduced. The second result, which is more important, holds under the following sign condition:

$$(1.10)\qquad f \geq 0 \quad \text{a.e. in } \Omega,$$

and

$$(1.11)\qquad y_d \leq 0 \quad \text{a.e. in } \Omega.$$

THEOREM 1.1. (a) <u>Under the assumptions (1.1), (1.2) and (1.6) for any optimal pair (c,y), there exists p satisfying the system</u>

$$-\Delta y + \lambda y + cy = f \quad \text{a.e. in } \Omega, \ y \in W^{2,2}(\Omega) \cap H_0^1(\Omega) \cap L^\infty(\Omega),$$

(1.12)
$$-\Delta p + \lambda p + cp - y = -y_d \quad \text{a.e. in } \Omega, \ p \in W^{2,2}(\Omega) \cap H_0^1(\Omega) \cap L^\infty(\Omega),$$

$$c = \frac{1}{N}py \quad \text{a.e. in } \Omega \cap \{c>0\},$$

$$py \leq 0 \quad \text{a.e. in } \Omega \cap \{c=0\}.$$

(b) Under the assumptions (1.1), (1.2), (1.6), (1.10) and (1.11) for any optimal pair (c,y), there exists p satisfying the following nonlinear elliptic system.

(1.13)
$$-\Delta y + \lambda y + \frac{1}{N}p\, y^2 = f \quad \text{a.e. in } \Omega, \ y = 0 \text{ at } \partial\Omega, \ y \in W^{2,q}(\Omega),$$
$$-\Delta p + \lambda p + \frac{1}{N}\, y\, p^2 - y = -y_d \quad \text{a.e. in } \Omega, \ p = 0 \text{ at } \partial\Omega, \ p \in W^{2,q}(\Omega),$$

and $c = \frac{1}{N}py$. Moreover, $y \geq 0$ and $p \geq 0$.

In the rest of this section we specialize for the case when (1.10) and (1.11) hold. Also, we assume that

(1.14) λ is sufficiently large.

REMARK 1.2. From (1.5), (1.6) and (1.13) we see that there is a uniform constant C_0, depending only on data (moreover, independent of λ and N), such that, for any optimal pair (c,y) and any corresponding p, we have

(1.15) $$\max \left\{ \|y\|_{L^\infty(\Omega)}, \ \|p\|_{L^\infty(\Omega)} \right\} \leq C_0.$$

Then, we can write (1.14) explicitly, for example, as

(1.14)* $$\lambda \geq \frac{1}{2} \left| \frac{2}{N}(C_0)^2 - 1 \right|.$$

PROPOSITION 1.2. Under the assumptions (1.1), (1.6), (1.10), (1.11), and (1.14) there may be just one positive solution of (1.13).

We now proceed to construct a solution.

Start with a simple problem:

(1.16)
$$\begin{cases} \text{Find } u \in W^{2,q}(\Omega),\ u \geq 0, \text{ such that,} \\ \qquad -\Delta u + \lambda u + bu^2 = f \quad \text{a.e. in } \Omega, \\ \qquad u = 0 \quad \text{at } \partial\Omega. \end{cases}$$

We assume that

(1.17) $b \in L_+^\infty(\Omega)$.

Then we have the following

PROPOSITION 1.3. Under the assumptions (1.1), (1.2), (1.10), and (1.17) there exists a unique solution of the problem (1.16).

The proof is based on the following iterative scheme. Let u_0 be defined as a solution of

$$(1.18)\qquad \begin{aligned} -\Delta u_0 + \lambda u_0 &= f \quad \text{a.e. in } \Omega, \\ u_0 &= 0 \text{ at } \partial\Omega. \end{aligned}$$

Observe that (1.10) implies that $u_0 \geq 0$. Now, choose constant $M \geq 0$, such that

$$(1.19)\qquad \begin{aligned} &-\|b\|_{L^\infty(\Omega)} v^2 + Mv \\ &\text{is an increasing function of } v, \text{ for } v \in \left[0, \|u_0\|_{L^\infty(\Omega)}\right]. \end{aligned}$$

Next, for $k \geq 1$, assuming we already have u_{k-1}, we define u_k as a solution of

$$(1.20)\qquad \begin{aligned} -\Delta u_k + (\lambda+M)u_k &= f - b(u_{k-1})^2 + Mu_{k-1} \quad \text{a.e. in } \Omega, \\ u_k &= 0 \quad \text{at } \partial\Omega. \end{aligned}$$

Then, one can show that the sequence $\{u_k\}$ converges to the unique positive solution of (1.16).

Next, we study the following (intermediate) problem:

$$(1.21)\qquad \begin{cases} \text{Find } (y,p) \in (W^{2,q}(\Omega))^2, \; y \geq 0, \; p \geq 0, \text{ such that} \\ \quad -\Delta y + \lambda y + \frac{1}{N} p\, y^2 = f \quad \text{a.e. in } \Omega, \; y = 0 \quad \text{at } \partial\Omega, \\ \quad -\Delta p + \lambda p + \frac{1}{N} y\, p^2 = g \quad \text{a.e. in } \Omega, \; p = 0 \quad \text{at } \partial\Omega. \end{cases}$$

We shall assume that (for q as in (1.1))

(1.22) $\quad g \in L^q_+(\Omega)$.

We have the following

PROPOSITION 1.4. <u>Under the assumptions (1.1), (1.2), (1.10), and (1.22) there exists a solution (y,p) of (1.21, with the following comparison property</u>:

(1.23) $\quad g^* \leq g \Rightarrow p^* \leq p, \; y^* \geq y$.

The proof is based on the following iterative scheme. Let y_0 be defined as a solution of

(1.24) $\quad -\Delta y + \lambda y = f \quad \text{a.e. in } \Omega, \; y = 0 \text{ at } \partial\Omega$.

Next, using Proposition 1.3, define p_1 as a solution of

(1.25) $\quad -\Delta p + \lambda p + \frac{1}{N} y_1 p^2 = g \quad \text{a.e. in } \Omega, \; p = 0 \text{ at } \partial\Omega$,

Further, define y_2 as a solution of

(1.26) $\quad -\Delta y + \lambda y + \frac{1}{N} p_1 y^2 = f \quad \text{a.e. in } \Omega, \; y = 0 \text{ at } \partial\Omega$,

and so on. It turns out that the sequence $\{(y_k, p_k)\}$ converges to the solution of (1.21).

Finally, consider the iterative scheme which solves the optimality system:

Let y_0 be defined as a solution of

(1.27) $$-\Delta y_0 + \lambda y_0 = f \quad \text{a.e. in } \Omega,\ y_0 = 0 \text{ at } \partial\Omega.$$

Define (y_k, p_k), for $k \geq 1$, as a solution, constructed in Proposition 1.4, of

(1.28) $$\begin{aligned} &-\Delta y_k + \lambda y_k + \tfrac{1}{N} p_k y_k^2 = f \quad \text{a.e. in } \Omega,\ y_k = 0 \text{ at } \partial\Omega, \\ &-\Delta p_k + \lambda p_k + \tfrac{1}{N} y_k p_k^2 = y_{k-1} - y_d \quad \text{a.e. in } \Omega,\ p_k = 0 \text{ at } \partial\Omega. \end{aligned}$$

THEOREM 1.2. Under the assumptions (1.1), (1.6), (1.10), (1.11) and (1.14) (or (1.14)*), the following holds:

(1.29) $$y_k \to y,\ p_k \to p, \text{ as } k \to \infty,$$

weakly in $W^{2,q}(\Omega)$, for a full sequence. Hence, (y,p) is the unique positive solution of (1.13), and $c_0 \equiv \frac{1}{N} py$ is the unique optimal control. Moreover, the following holds:

(1.30) $$y_{2k} \searrow y,\ y_{2k+1} \nearrow y,\ p_{2k} \nearrow p,\ p_{2k+1} \searrow p, \text{ as } k \to \infty.$$

2. OPTIMAL IMPLICIT DAMPING CONTROL. Let

(2.1) $$d \in L_+^\infty(\Omega),$$

and

(2.2) $$g \in L^q(\Omega),\ q \geq 2,\ q > \tfrac{n}{2},$$

be given. For any $f \in L_+^2(\Omega)$, we define $v = v(f)$, as a solution of the following equation

(2.3) $$\begin{aligned} &-\Delta v + \lambda v + d[(-\Delta+\lambda)^{-1} f]v = g \quad \text{a.e. in } \Omega, \\ &\quad v \in W^{2,2}(\Omega) \cap H_0^1(\Omega) \cap L^\infty(\Omega), \\ &\quad (-\Delta+\lambda)^{-1} f \in W^{2,2}(\Omega) \cap H_0^1(\Omega), \end{aligned}$$

or more explicitly, we define $(u,v) = u((f),v(f))$ as a solution of the following mildly nonlinear system

(2.4) $$\begin{aligned} &-\Delta u + \lambda u = f \quad \text{a.e. in } \Omega, \\ &\quad u \in W^{2,2}(\Omega) \cap H_0^1(\Omega), \\ &-\Delta v + \lambda v + duv = g \quad \text{a.e. in } \Omega, \\ &v \in W^{2,2}(\Omega) \cap H_0^1(\Omega) \cap L^\infty(\Omega). \end{aligned}$$

It is easy to see that (2.4) has the unique solution. Also,

(2.5) $$\begin{cases} \|v(f)\|_{L^\infty(\Omega)} \leq \text{const.} \\ \text{const. does not depend on } \lambda \in R_+ \text{ and } f \in L_+^2(\Omega). \end{cases}$$

Next, for (q as in (2.2))

(2.6) $\qquad v_d \in L^q(\Omega)$

and $N > 0$ given, we define the cost functional J by

(2.7) $\qquad J(f) = \frac{1}{2} \int_\Omega (v(f) - v_d)^2 dx + \frac{N}{2} \int_\Omega f^2 dx.$

As before we are looking for an $f_0 \in L^2_+(\Omega)$, such that

(2.8) $\qquad J(f_0) = \min_{f \in L^2_+(\Omega)} J(f).$

Similarly as before, we have

PROPOSITION 2.1. <u>There exists an optimal control.</u>

Again, for strong results we shall need a sign condition:

(2.9) $\qquad g \geq 0$ a.e. in Ω,

(2.10) $\qquad v_d \leq 0$ a.e. in Ω.

THEOREM 2.1. (a) <u>Under the assumptions (1.2), (2.1), (2.2), and (2.6), for any optimal control f, and associated solution of the state equation (u,v), there exists</u> (p_1, p_2) <u>satisfying the system</u>

(2.11)
$$\begin{aligned}
&-\Delta u + \lambda u = f \quad \text{a.e. in } \Omega, \ u \in W^{2,2}(\Omega) \cap H^1_0(\Omega),\\
&-\Delta v + \lambda v + duv = g \quad \text{a.e. in } \Omega, \ v \in W^{2,2}(\Omega) \cap H^1_0(\Omega) \cap L^\infty(\Omega),\\
&-\Delta p_1 + \lambda p_1 - dvp_2 = 0 \quad \text{a.e. in } \Omega, \ p_1 \in W^{2,q'}(\Omega), \ p_1 = 0 \text{ at } \partial\Omega,\\
&-\Delta p_2 + \lambda p_2 + dup_2 - v = -v_d \quad \text{a.e. in } \Omega, \ p_2 \in W^{2,2}(\Omega) \cap H^1_0(\Omega) \cap L^\infty(\Omega),\\
&f = \tfrac{1}{N} p_1 \quad \text{a.e. in } \Omega \cap \{f>0\},\\
&p_1 < 0 \quad \text{a.e. in } \Omega \cap \{f=0\},
\end{aligned}$$

<u>for any</u> $q' < \infty$.

(b) <u>Under the assumptions (1.2), (2.1), (2.2), (2.6), (2.9) and (2.10), for any optimal control f, and associated solution to the state equation (u,v), there exists</u> (p_1, p_2) <u>satisfying the following nonlinear elliptic system</u>

(2.12)
$$\begin{aligned}
&-\Delta u + \lambda u - \tfrac{1}{N} p_1 = 0 \quad \text{a.e. in } \Omega, \ u \in W^{2,q'}(\Omega), \ u = 0 \text{ at } \partial\Omega,\\
&-\Delta v + \lambda v + duv = g \quad \text{a.e. in } \Omega, \ v \in W^{2,q}(\Omega), \ v = 0 \text{ at } \partial\Omega,\\
&-\Delta p_1 + \lambda p_1 - dvp_2 = 0 \quad \text{a.e. in } \Omega, \ p_1 \in W^{2,q'}(\Omega), \ p_1 = 0 \text{ at } \partial\Omega,\\
&-\Delta p_2 + \lambda p_2 + dup_2 - v = -v_d \quad \text{a.e. in } \Omega, \ p_2 \in W^{2,q}(\Omega), \ p_1 = 0 \text{ at } \partial\Omega,
\end{aligned}$$

<u>for any</u> $q' < \infty$, <u>and</u> $f = \frac{1}{N} p_1$. <u>Moreover</u>, $u \geq 0$, $v \geq 0$, $p_1 \geq 0$ and $p_2 \geq 0$.

PROPOSITION 2.2. <u>Under the assumptions (1.2), (2.1), (2.2), (2.6), (2.10) (2.11) and (1.14), there may be just one bounded positive solution of (2.11).</u>

We now proceed to construct a solution of the optimality system. Consider the following auxiliary problem. Given $h \in L^{\infty}_{+}(\Omega)$, find $(u,v,p_1,p_2) \in W^{2,q'}(\Omega)\times W^{2,q}(\Omega)\times W^{2,q'}(\Omega)\times W^{2,1}(\Omega)$ $\forall q' < \infty$, such that

$$
(2.11)\quad \begin{aligned}
&-\Delta u + \lambda u = h \quad \text{a.e. in } \Omega,\ u = 0 \text{ at } \partial\Omega,\\
&-\Delta v + \lambda v + duv = g \quad \text{a.e. in } \Omega,\ v = 0 \text{ at } \partial\Omega,\\
&-\Delta p_1 + \lambda p_1 - dvp_2 = 0 \quad \text{a.e. in } \Omega,\ p_1 = 0 \text{ at } \partial\Omega,\\
&-\Delta p_2 + \lambda p_2 + dup_2 - v = -v_d \quad \text{a.e. in } \Omega,\ p_2 = 0 \text{ at } \partial\Omega.
\end{aligned}
$$

This is a trivial system. Indeed, $u \Rightarrow v \Rightarrow p_2 \Rightarrow p_1$. Also, we observe that

$$
(2.12)\quad \bar{h} \ge h \Rightarrow \bar{u} \ge u \Rightarrow \bar{v} \le v \Rightarrow \bar{p}_2 \le p_2 \Rightarrow \bar{p}_1 \le p_1.
$$

We are now ready to solve the original problem. Consider the following iterative scheme. Let $(u_0,v_0,p_{1,0},p_{2,0})$ be a solution of $u = 0$ in Ω,

$$
(2.13)\quad \begin{aligned}
&-\Delta v + \lambda v = g \quad \text{a.e. in } \Omega,\ v = 0 \text{ at } \partial\Omega,\\
&-\Delta p_1 + \lambda p_1 - dvp_2 = 0 \quad \text{a.e. in } \Omega,\ p_1 = 0 \text{ at } \partial\Omega,\\
&-\Delta p_2 + \lambda p_2 + dup_2 - v = -v_d \quad \text{a.e. in } \Omega,\ p_2 = 0 \text{ at } \partial\Omega.
\end{aligned}
$$

Define, by induction, for $k \ge 1$, $(u_k,v_k,p_{1,k},p_{2,k})$ as a solution of

$$
(2.14)\quad \begin{aligned}
&-\Delta u + \lambda u = \tfrac{1}{N}p_{1,k-1} \quad \text{a.e. in } \Omega,\ u = 0 \text{ at } \partial\Omega,\\
&-\Delta v + \lambda v + duv = g \quad \text{a.e. in } \Omega,\ v = 0 \text{ at } \partial\Omega,\\
&-\Delta p_1 + \lambda p_1 - dvp_2 = 0 \quad \text{a.e. in } \Omega,\ p_1 = 0 \text{ at } \partial\Omega,\\
&-\Delta p_2 + \lambda p_2 + dup_2 - v = -v_d \quad \text{a.e. in } \Omega,\ p_2 = 0 \text{ at } \partial\Omega.
\end{aligned}
$$

Then we have the following

THEOREM 2.2. *Under the assumptions (1.2), (2.1), (2.2), (2.6), (2.9), (2.10) and (1.14), the following holds:*

$$
(2.15)\quad u_k \to u,\ v_k \to v,\ p_{1,k} \to p_1,\ p_{2,k} \to p_2, \text{ as } k \to \infty,
$$

weakly in $W^{2,q}(\Omega)$, for a full sequence. Hence, (u,v,p_1,p_2) is the unique positive solution of (2.11) and $f = \frac{1}{N}p_1$ is the unique optimal control. Moreover, the following holds:

$$
(2.16)\quad \begin{aligned}
&u_{2k} \nearrow u,\ u_{2k+1} \searrow u,\ v_{2k} \searrow v,\ v_{2k+1} \nearrow v,\\
&p_{1,2k} \nearrow p_1,\ p_{1,2k+1} \searrow p_1,\ p_{2,2k} \searrow p_{2,2k+1} \nearrow p_2.
\end{aligned}
$$

as $k \to \infty$.

PROOF. It is not difficult to conclude from (2.13) that

$$
(2.17)\quad \begin{aligned}
&u_{2k} \nearrow,\ u_{2k+1} \searrow,\ v_{2k} \searrow,\ v_{2k+1} \nearrow,\\
&p_{1,2k} \nearrow,\ p_{1,2k+1} \searrow,\ p_{2,2k} \searrow,\ p_{2,2k+1} \nearrow.
\end{aligned}
$$

Hence, there exists a $(\underline{u},\bar{v},\underline{p}_1,\bar{p}_2,\bar{u},\underline{v},\bar{p}_1,\underline{p}_2)$, such that

$$u_{2k} \nearrow \underline{u},\ u_{2k+1} \searrow \bar{u},\ v_{2k} \searrow \bar{v},\ v_{2k+1} \nearrow \underline{v},$$
(2.18)
$$p_{1,2k} \nearrow \underline{p}_1,\ p_{1,2k+1} \searrow \bar{p}_1,\ p_{2,2k} \searrow \bar{p}_2,\ p_{2,2k+1} \nearrow \underline{p}_2,$$

and also, by elliptic estimates, weakly in $W^{2,q}(\Omega)$ (actually, u_1's and $p_{1,1}$'s converge weakly in $W^{2,q'}(\Omega)\ \forall_{q'} < \infty$). Furthermore, we have, for $k \geq 1$,

$$-\Delta u_{2k} + \lambda u_{2k} - \frac{1}{N}p_{1,2k-1} = 0 \quad \text{a.e. in } \Omega,\ u_{2k} = 0 \text{ at } \partial\Omega,$$
$$-\Delta v_{2k} + \lambda v_{2k} + du_{2k}v_{2k} = g \quad \text{a.e. in } \Omega,\ v_{2k} = 0 \text{ at } \partial\Omega,$$
$$-\Delta p_{1,2k} + \lambda p_{1,2k} - dv_{2k}p_{2,2k} = 0 \quad \text{a.e. in } \Omega,\ p_{1,2k} = 0 \text{ at } \partial\Omega,$$
$$-\Delta p_{2,2k} + \lambda p_{2,2k} + dup_{2,2k} - v_{2k} = -v_d \quad \text{a.e. in } \Omega,\ p_{2,2k} = 0 \text{ at } \partial\Omega.$$

and

$$-\Delta u_{2k+1} + \lambda u_{2k+1} - \frac{1}{N}p_{1,2k} = 0 \quad \text{a.e. in } \Omega,\ u_{2k+1} = 0 \text{ at } \partial\Omega,$$
$$-\Delta v_{2k+1} + \lambda v_{2k+1} + du_{2k+1}v_{2k+1} = g \quad \text{a.e. in } \Omega,\ v_{2k+1} = 0 \text{ at } \partial\Omega,$$
$$-\Delta p_{1,2k+1} + \lambda p_{1,2k+1} - dv_{2k+1}p_{2,2k+1} = 0 \text{ a.e. in } \Omega,\ p_{1,2k+1} = 0 \text{ at } \partial\Omega,$$
$$-\Delta p_{2,2k+1} + \lambda p_{2,2k+1} + dup_{2,2k+1} - v_{2k+1} = -v_d \text{ a.e. in } \Omega,\ p_{2,2k+1} = 0 \text{ at } \partial\Omega.$$

Passing the limit, we conclude that $(\underline{u},\bar{v},\underline{p}_1,\bar{p}_2,\bar{u},\underline{v},\bar{p}_1,\underline{p}_2)$ is a solution of

$$-\Delta\underline{u} + \lambda\underline{u} - \frac{1}{N}\bar{p}_1 = 0 \quad \text{a.e. in } \Omega,\ \underline{u} = 0 \text{ at } \partial\Omega,$$
$$-\Delta\bar{v} + \lambda\bar{v} + d\underline{u}\bar{v} = g \quad \text{a.e. in } \Omega,\ \bar{v} = 0 \text{ at } \partial\Omega,$$
$$-\Delta\underline{p}_1 + \lambda\underline{p}_1 - d\bar{v}\bar{p}_2 = 0 \quad \text{a.e. in } \Omega,\ \underline{p}_1 = 0 \text{ at } \partial\Omega,$$
(2.19)
$$-\Delta\bar{p}_2 + \lambda\bar{p}_2 + d\underline{u}\bar{p}_2 - \bar{v} = -v_d \quad \text{a.e. in } \Omega,\ \bar{p}_2 = 0 \text{ at } \partial\Omega,$$
$$-\Delta\bar{u} + \lambda\bar{u} - \frac{1}{N}\underline{p}_1 = 0 \quad \text{a.e. in } \Omega,\ \bar{u} = 0 \text{ at } \partial\Omega,$$
$$-\Delta\underline{v} + \lambda\underline{v} + d\bar{u}\underline{v} = g \quad \text{a.e. in } \Omega,\ \underline{v} = 0 \text{ at } \partial\Omega,$$
$$-\Delta\bar{p}_1 + \lambda\bar{p}_1 - d\underline{v}\underline{p}_2 = 0 \quad \text{a.e. in } \Omega,\ \bar{p}_1 = 0 \text{ at } \partial\Omega,$$
$$-\Delta\underline{p}_2 + \lambda\underline{p}_2 + d\bar{u}\underline{p}_2 - \underline{v} = -\underline{v}_d \quad \text{a.e. in } \Omega,\ \underline{p}_2 = 0 \text{ at } \partial\Omega.$$

Then, by inspection, we see that $(\bar{u},\underline{v},\bar{p}_1,\underline{p}_2,\underline{u},\bar{v},\underline{p}_1,\bar{p}_2)$ is a solution, as well. But, quite similarly as in the Proposition 2.2, we can show that there may be just one solution of (2.19). Hence,

$$(\underline{u},\bar{v},\underline{p}_1,\bar{p}_2,\bar{u},\underline{v},\bar{p}_1,\underline{p}_2) = (\bar{u},\underline{v},\bar{p}_1,\underline{p}_2,\underline{u},\bar{v},\underline{p}_1,\bar{p}_2),$$

and we deduce (2.16). The Theorem follows now easily.

REFERENCES

1. V. Barbu, Optimal control of variational inequalities, Research Notes in Mathematics 100, Pitman, London, 1984.

2. A. Friedman, Nonlinear optimal control problems for parabolic equations, SIAM J. Control & Optimiz., 22, 805-816, 1984.

3. D. Gilbarg and N.S. Trudinger, Elliptic partial differential equations of second order, Second Ed., Springer-verlag, Berlin, 1983.

4. A. Leung, Systems of nonlinear partial differential equations and applications, Kluwer Academic Publishers, Dordrecht, 1989.

5. J.L. Lions, Optimal control of systems governed by partial differential equations, Springer-Verlag, Berlin, 1971.

6. F. Mignot and J.P. Puel, Optimal control in some variational inequalities, SIAM J. Control & Optimiz., 22, 466-476, 1984.

7. S. Stojanovic, Optimal damping control and nonlinear parabolic systems, Numerical Functional Analysis & Optimiz., 10(586), 573-591, 1989.

DEPARTMENT OF MATHEMATICAL SCIENCES
UNIVERSITY OF CINCINNATI
CINCINNATI, OHIO 45221-0025